AFTERSHOCKS OF DISASTER

AFTERSHOCKS
OF DISASTER

Puerto Rico Before and After the Storm

Yarimar Bonilla and Marisol LeBrón

Haymarket Books
Chicago, Illinois

Published in 2019 by
Haymarket Books
P.O. Box 180165
Chicago, IL 60618
773-583-7884
www.haymarketbooks.org
info@haymarketbooks.org

ISBN: 978-1-64259-030-2

Distributed to the trade in the US through Consortium Book Sales
and Distribution (www.cbsd.com) and internationally through Ingram Publisher Services International (www.ingramcontent.com).

This book was published with the generous support of Lannan
Foundation and Wallace Action Fund.

Special discounts are available for bulk purchases by organizations
and institutions. Please call 773-583-7884 or email
info@haymarketbooks.org for more information.

Cover artwork, Pentagramas (Staves) series, detail, 2012, 18 x 12 in,
paper cutout, by Frances Gallardo
Cover design by Rachel Cohen.

Printed in Canada by union labor.

Library of Congress Cataloging-in-Publication data is available.

10 9 8 7 6 5 4 3 2

TABLE OF CONTENTS

PART III: REPRESENTING THE DISASTER

PART IV: CAPITALIZING ON THE CRISIS

FOREWORD

Arcadio Díaz-Quiñones

Long before Hurricane María, the fiscal and political crisis had invested all things Puerto Rican with increasing urgency. The threat of total collapse continues to raise major themes of discussion in a global context: colonial capitalism, human rights, gender equality, democracy, unpayable debt, climate change, migration and citizenship, environmental policies, education, and health care. And yet for many, the newly discovered US "territory" and the millions of second-class American citizens remain mysterious, or rather invisible, right before our eyes. Fortunately, there are signs that Puerto Rico's and the United States' long, transnational history—often silenced in the American mainstream—now commands serious attention from those trying to come to a better understanding of the legacies of colonialism and of Puerto Rican resistance.

The willingness to ask a huge array of new questions inspired the organizers of the conference entitled "Aftershocks of Disaster—Puerto Rico a Year after María," held at Rutgers University in New Brunswick, New Jersey, in September 2018. I attended the Rutgers colloquium, and what I saw and listened to in the give-and-take of conversation that day has stayed with me. The fact that this event was held in New Jersey—and that other colloquia animated by a similar spirit have been held in New York, Massachusetts, Connecticut, Illinois, and Washington, DC—shows, first, the strength of the ethos of solidarity among diverse diasporic communities and the institutions they have created as well as the moral sensibility of their allies at universities and research centers. Second, and equally important, it shows how the full extent of the disaster is driving the research, writing, and activism of a new generation of scholars and journalists. This is all particularly encouraging, especially in view of the drastic budgetary cuts that are proving so destructive for the Universidad de Puerto Rico at the worst possible time.

Like other islands in the Caribbean, current struggles in Puerto Rico have roots deep in colonial history, dating back to the old Spanish colony and the military occupation by the United States in 1898. Still, even in academia, some have only the vaguest idea of that history, despite the crucial fact that the Puerto Rican population in the United States—a story spanning more than a century—has grown to well over five million. History seems to have been erased from memory. Few remember, for instance, the deaths of thousands of Puerto Rican soldiers in the Korean and the Vietnam Wars. The United States' imperial domination is rarely acknowledged despite significant scholarly and journalistic publications in Spanish and English, and of the revelatory poetic voices of Julia de Burgos, Pedro Pietri, and so many others down to the present day.

The political, conceptual, and emotional impact of earthshaking transformations in Puerto Rico is far-reaching. As I listened to the compelling presentations at the Rutgers meeting, it soon became clear that the aftermath of María—as has also been said of Hurricane Katrina—reveals much that was concealed. This is true not only of the extreme social inequalities or the defunding of public education but also of the complicity of specific local Puerto Rican economic and political actors with neoliberal policies. At the same time, the crisis has given new visibility to a powerful and racist imperial state and to the repression of radical Puerto Rican voices, a history that has left indelible long-term effects on many lives.

There are still euphemisms all over the place to cover the fact that Puerto Rico has never had full control of its economic, environmental, and communications policies. But thanks to the amazing investigative journalism and work by activists, some of whom were present at Rutgers, it has been sufficiently revealed that enormous problems had been swirling for years just below the surface. And that after the turmoil of the debt crisis, the passing of the Financial Control and Management Board Stability Act (known on the island as PROMESA) in 2016 by the US Congress, and then Hurricane María, there does not seem to be much left of the Estado Libre Asociado, or the Commonwealth of Puerto Rico (both names are quite misleading). The imposition of PROMESA and of the board (popularly known as La Junta) constitute a decisive rebuttal to those who believed that the Estado Libre Asociado, created in 1952, represented the decolonization of Puerto Rico.

But perhaps it is no less true that disasters keep concealing as much as they reveal. Such is the case with the decision faced by many Puerto Ricans whether to stay or to leave the island. A new and massive exodus to the United States, often of younger workers and professionals—as well as families with children—has come into view. Indeed, since the mid-twentieth century, the San Juan airport has acquired immense material and symbolic power. It is a central site of memory. But in terms of current political mobilization in the struggle for equal rights, what are the gains and losses of leaving? To put this question is also to ask, What challenges lie ahead for those who remain behind? Are their voices weakened or strengthened? One should also take note of a change in the lived experience shared by many in the diaspora and on the island. The magnitude of the crisis has brought a transformative personal, political, and intellectual relationship with individuals and activist groups, perhaps a new understanding of *belonging*. It often includes a keen awareness of a larger and diverse community and a bilingual translating relationship, into English or Spanish. All of this shows, in a crucial respect, a new way of seeing and feeling that challenges conventional definitions of identity. I keep thinking of the questions and perspectives opened up by Albert O. Hirschman's reflections on the ethical and political dilemmas raised by *Exit, Voice, and Loyalty*.

But something else, and most important, is happening on the ground as Puerto Ricans struggle to assert their dignity. There is much to learn from how many have found spiritual strength to cope with the ravages of disaster capitalism, reimagining themselves in the face of traumatic loss of thousands of lives and exploring their own vulnerabilities. Community leaders are yearning not for epic confrontations but for a new sense of self and community and for new political beginnings. Journalists, writers, and academics are painstakingly describing how Puerto Ricans are transforming their society in the long shadow of disasters. The scholars and journalists who spoke at Rutgers—and others who have contributed essays for this volume—have eagerly taken up the task of bringing dark truths of economic, social, and environmental damage to light. They also bring critical perspectives to a new memory of dispossession and a vision for the future.

Who gets to tell the stories that need to be told? As a deeply committed public intellectual, Yarimar Bonilla deserves the credit for imagining and organizing the "Aftershocks" conference. She also deserves our gratitude. Her breadth of vision, command of the issues, and firsthand knowledge of academics and activists allowed her to bring together a wide range of performance and visual artists, writers, journalists, photographers, academics, documentary filmmakers, anthropologists, environmentalists, and legal scholars. These are "voices that need to be heard," as Bonilla said in her opening remarks.

The Rutgers meeting succeeded in creating a space for meaningful dialogue and probing analyses of the consequences of disasters and political disempowerment in Puerto Rico. It also illuminated multiple forms of resistance. The word *aftershocks* became a rich metaphor, with many shades of meaning. The dialogue was set in motion by two wonderful beginnings. The first was the performance by a theater group from Puerto Rico; the second was a memorable keynote conversation between Bonilla and Naomi Klein.

The play *¡Ay María!* provided a moving and scathingly funny window onto posthurricane lives. The characters circled each other and crossed paths in a mosquito net, diluting the boundaries between the public and the private. The interactive play captures the climate of frustration and the rage against state agencies' failures to even count the dead, and the imperial cynicism of President Trump tossing out paper towels. *¡Ay María!* shows how vulnerabilities and hopes are embedded in a distinctive political and cultural context in which beliefs, music, and dance can mobilize the community. The characters are always in motion, telling stories, in conversation with the audience, creating awareness of the present situation. Their movements and voices, working around the personal and the local, were really speaking about justice and democracy.

I was extraordinarily impressed by the dialogue between Bonilla and the distinguished political analyst Klein. Not only did they offer a deeply thoughtful assessment of the aftershocks' impact on Puerto Rico, but they also posed essential questions for further investigation. Furthermore, in their commentaries, they spoke with feeling and fullness of heart, remembering colleagues, friends, other conversations, and their own political and affective experiences on the island, thus setting the tone for the entire conference. It is clear that Klein and Bonilla share the conviction that disasters open up new possibilities for critical thinking and for art as a form of intervening in politics and imagining a radically different society. Their words continue to reverberate and inspire. We need that inspiration more than ever.

INTRODUCTION

Aftershocks of Disaster

Yarimar Bonilla and Marisol LeBrón

"There's no way to win. Those who stayed are suffering because of the situation back home. But those who left are suffering because of the circumstances under which they were forced to flee," explained Isabel, a thirty-year-old woman, to journalist Andrea González-Ramírez six months after Hurricane María devastated Puerto Rico.[1] The flooding caused by the hurricane's intense rainfalls forced Isabel, her husband, and her two young children to take refuge in a local shelter. Isabel and her family spent the next ten days in that shelter in Toa Baja with others whose homes had been severely damaged or destroyed during the storm. After spending more than a week in the emergency shelter, Isabel and her family, like thousands of others in the coming weeks and months, decided to relocate to the United States. They didn't step foot in their house again before migrating to Florida, leaving all their belongings behind. Although the family hoped that getting away from the storm's devastation would make their lives easier, Isabel soon started to experience serious bouts of depression. The family moved a second time, this time to Arizona, hoping that Isabel's depression might improve in a new location. But things only worsened for Isabel as she started to experience panic attacks. The thought of never being able to return to Puerto Rico seemed to be the biggest trigger for Isabel's growing anxiety. "I left thinking that we would be able to return. But there is no power, there are no jobs, so the exodus keeps extending. Accepting that I will never go back is one of the reasons that I'm facing issues with my mental health. We had the unrealistic expectation that life would normalize in the island, but that's not the case," Isabel said.[2]

Isabel's story provides a poignant example of how natural disasters do more than just mar the landscape; they upend people's lives, lingering and reverberating long after the winds have died down and the waters have calmed. In Isabel's story, we see that the trauma she describes only worsened with time. Her story complicates a linear timeline of disaster and recovery and points instead to natural disasters as cumulative and ongoing. Isabel's anxiety stems from a realization that her life will never be the same—that the Puerto Rico she knew will never be the same—but the storm accounts only in part for that feeling of loss. Isabel laments that there are no jobs and that more and more Puerto Ricans are leaving the island. These are realities that long predated the storm but were worsened by its impact. Isabel suggests that her expectation that things would get back to normal soon after the storm was unrealistic, but this only raises the question whether the social problems that plagued Puerto Rico before María aren't also part of the disaster causing Isabel's breath to shorten and her chest to tighten.

The concept of aftershocks is mostly used in the context of earthquakes to describe the jolts felt after the initial quake. Aftershocks can continue for days, weeks, months, and even years after the "mainshock." The bigger the earthquake, the more numerous and long-lasting the aftershocks will be. Although aftershocks are often smaller, their effects can compound the damage of the initial shock and create new urgencies that complicate recovery efforts.

Most of what is discussed in this book examines the aftershocks of Hurricane María, not just the effects of the wind or rain but also what followed: state failure, social abandonment, capitalization on human misery, and the collective trauma produced by the botched response. In the nearly two years since Hurricane María made landfall, Puerto Ricans have found themselves relentlessly jolted by the storm's aftershocks. This happens every time systemic failures are revealed, death and damages are denied, aid is refused, profiteering is discovered, and officials who were not elected by local residents make drastic decisions about the island's future. Much as we see in Isabel's case, these small but ongoing blows can have major repercussions that are worse and potentially more destructive than the initial event.

Aftershocks remind us that disasters are not singular events but ongoing processes. Building on this idea, *Aftershocks of Disaster* examines both Hurricane María's aftershocks and its foreshocks—the sociohistorical context of debt crisis, migration, and coloniality in which the storm took place. Indeed, we ask whether Hurricane María should be considered the "mainshock" at all, or whether the storm and its effects are best understood as the compounded results of a longer colonial history.

~&

On September 20, 2017, Hurricane María sliced through Puerto Rico, producing one of the deadliest natural disasters in US history and bringing unprecedented attention to this colonial territory. For many in Puerto Rico, as well as for those following the news from afar, one of the most searing memories of the storm's immediate aftermath is the press conference held by US president Donald Trump in which he bragged about the storm's low death count. According to Trump, Hurricane María was not a "real catastrophe" like Hurricane Katrina, which battered Louisiana and the Gulf Coast in 2005 and left eighteen hundred people dead. He boasted that in the case of María only sixteen people had died thanks to the preparation and performance of both the local and federal government. The official death tally began creeping up immediately after his visit, eventually stalling at sixty-four. But reporters, public health officials, funeral home directors, and Puerto Ricans who had lost someone as a result of the storm all maintained that the true number had to be much higher, given what they had witnessed and experienced. As Carla Minet documents in her contribution to this volume, journalists from both local and national outlets partnered together and, along with lawyers and independent researchers, began revealing the truth of María's fatal consequences. In the end, the local government accepted 2,975 as the official death toll, even though some studies place it much higher and there has yet to be a definitive accounting.

On September 12, 2018, only a few days before the first anniversary of the storm's landfall in Puerto Rico, Trump tweeted that the revised death toll was little more than a partisan conspiracy theory. Local politicians, for their part, have accepted the higher numbers but refuse to account for their mishandling of death certifications, reported cremations without autopsies, and the broader forms of structural negligence that caused the deaths in the first place. We now know that the great majority of those who lost their lives to María perished not because of the storm but because of the structural failures that followed it: uncleared roads that did not allow ambulances to arrive, lack of water distribution that led residents to contaminated water sources, lack of generators in hospitals, and more than half a year without electricity to power medical equipment, refrigerate lifesaving medications such as insulin, and provide public lighting and traffic lights to prevent deadly accidents. Lives were not lost to the wind and the rain, or even to Trump's disrespect; instead, residents drowned in bureaucracy and institutional neglect.

Further, in the weeks and months spent waiting for the power to come back on, or worrying about whether the water they were giving their children was safe to consume, or wondering if they were going to be able to go back to work or school, many Puerto Ricans were stricken with anxiety, fear, and a deep feeling of abandonment, which exacerbated an often hidden mental health crisis on the island. Suicides spiked after the storm, as did cases of domestic and intimate partner violence. Suicides increased 28 percent in 2017, while calls to suicide prevention hotlines doubled from September 2017 to March 2018.[3] Groups working with women and families experiencing domestic violence and intimate abuse also reported a surge in requests for services and preventive educational programs.[4] As journalist and activist Mari Mari Narváez notes in her interview with Marisol LeBrón in this volume, marginalized populations only found their vulnerability intensified as the state struggled to provide even the most basic resources and protections.

In this context, it is difficult to predict when the disaster associated with Hurricane María will actually end since each aftershock creates a new series of problems that ripple out far and wide. Nor is it even clear when this disaster began. Long before María, Puerto Rico was already suffering the effects of a prolonged economic recession, spiraling levels of debt, and deep austerity cuts to public resources. This was preceded by over five centuries of colonialism (first Spanish, then American) and a long history of structural vulnerability and forced dependency. For example, Puerto Rico's weakened local government is subject to the whims of Washington and thus unable to chart political and economic policy centered around local needs (a case in point being the inability to repeal the Jones Act, which requires all consumer goods to arrive on US vessels, drastically raising the cost of basic need items). Moreover, its social safety nets are constrained by the structural inequalities between the United States and its territories. All of this results in depressed wages, restricted social security and Medicare benefits, and a poverty rate that is more than twice the national average. These structural vulnerabilities, as contributors to this volume demonstrate, set the stage for María's impact.

COLONIAL DEBT AND DISASTER

Long before the hurricane, Puerto Rico felt to many like a society in ruin—financially and politically. During the past two decades, this US territory was plunged into a deep economic recession as tax incentives for foreign companies were phased out. Companies left in droves in search of less regulation and greater corporate welfare. Almost immediately unemployment started to increase, public coffers dwindled, residents were told to tighten their belts, and many began migrating in greater numbers in search of economic stability and opportunity.

Puerto Rico's public debt, which eventually grew to more than $72 billion, helped lay the groundwork that made Hurricane María so devastating and the recovery so slow. The debt ballooned as Puerto Rican officials turned to Wall Street to address the economic stagnation that followed corporate flight, increasingly taking on greater debt in an attempt to stay afloat.

Puerto Rico's debt crisis was also fueled by the particular financial apparatus of Puerto Rican bonds. In addition to having repayment guaranteed in the constitution, bonds issued by the Puerto Rican government also have the unique and singular quality (unavailable within any of the fifty states) of being triple tax-exempt—free of tax obligations at the state, federal, or local level. This made them irresistible to Wall Street financiers. When debt levels pushed beyond constitutional limits, new mechanisms of economic capture were created. A sales tax suggested by financial strategists at Lehman Brothers was implemented in 2006 to guarantee new loans. Meanwhile, public infrastructure (such as airports, bridges, and hospitals) was increasingly sold to the highest bidder, further eroding public coffers.[5] As a result about one-third of Puerto Rico's budget is now funneled toward servicing a debt that many believe is both unconstitutional and unsustainable.[6] Puerto Rico's contemporary debt crisis, however, is a symptom of a much deeper economic and political malaise stemming from its unresolved colonial status. After the establishment in 1952 of the Estado Libre Asociado (Freely Associated State, often glossed in English as the "commonwealth"), Puerto Rico was imagined to have "the best of both worlds": it had a semblance of local sovereignty, backed by the economic and political protections that came with ties to the United States.

During the mid-twentieth century, Puerto Rico went from being the "poorhouse of the Caribbean" to the "shining star" of US democracy in the region, as the island was rapidly industrialized and the standard of living increased for many Puerto Ricans. The gains seemingly achieved through Puerto Rico's commonwealth status caused many to ignore fundamental flaws in this political-economic arrangement. Puerto Rico's territorial status prevents the local government from shaping and implementing many of its own policies, encourages an overdependence on US capital investment, and hampers long-term and sustainable economic growth that would benefit the local population. This makes Puerto Rico particularly vulnerable during periods of economic contraction, and, indeed, for the past two decades, many residents have felt that they were already living in a state of crisis. By the time the debt crisis hit, local and federal policy makers had seemingly run out of options for staving off collapse.

In an effort to stop Puerto Rico's debt "death spiral," the local government declared the debt "unpayable" in 2016 and signed an emergency bill placing a moratorium on debt payments.[7] When the local government sought to declare bankruptcy, the real nature of Puerto Rico's limited sovereignty became clear. With the legal status of neither a state nor an independent nation, Puerto Rico could not refinance or default on its debt. The US Congress denied Puerto Rico not only the right to bankruptcy but also any kind of financial bail-out or meaningful redress. Instead, Congress "assisted" Puerto Rico by imposing the Puerto Rico Oversight, Management, and Economic Stability Act, or PROMESA bill, which established an unelected Fiscal Control Board to manage local finances and renegotiate the debt. The board is colloquially referred to as "la Junta" by locals, which signals many Puerto Ricans' perception of the board as a dictatorial body that has seized power from the local government.

The members of the Fiscal Control Board were appointed by the US Congress with virtually no local input and no form of local accountability, yet all costs are paid directly by Puerto Rican taxpayers to the tune of $200 million a year.[8] According to the current president of the Fiscal Control Board, José B. Carrión III, the fact that the board does not have to answer to either the Puerto Rican government or to local citizens is precisely what allows it to make unpopular choices that are necessary to improve Puerto Rico's economy.[9]

The board has no vision for the island's future other than restoring its ability to continue borrowing and generating profit for investors. It has focused solely on imposing structural-adjustment-style austerity measures, even after many international monetary institutions, including the International Monetary Fund, have admitted that these policies are shortsighted and doomed to fail.[10] The board's failure to invest in the well-being and livelihoods of local Puerto Ricans is evident in its initial targeting of public education—including the University of Puerto Rico, as Rima Brusi and Isar Godreau detail in their essay—and the imposition of drastic cuts to pensions and wages.

Given all this, Puerto Rico was already in a state of political and social crisis long before the winds of María arrived. Protests against PROMESA were taking place across the territory, student strikes had shut down the university for months, abandoned schools and foreclosed buildings were being taken over as community centers, and across the urban landscape new street art appeared calling for a reimagining of what decolonization and self-determination might look like for a bankrupt colony. Activists had already come together to tell the government "se acabaron las promesas" (the time for promises is over), a slogan that evokes both the PROMESA law and the decades of false promises that had followed the establishment of Puerto Rico's commonwealth status.

For young people in particular, the economic advancement and social freedoms promised by the commonwealth arrangement proved to be empty promises as insecurity, precarity, and vulnerability marked their lives and increasingly constrained their futures. They worried that their lives were essentially being mortgaged to service the debt and generate profits for vulture capitalists. Education, well-paying jobs, affordable housing, and the ability to build one's life in Puerto Rico were increasingly out of grasp for many Puerto Rican youth, who faced a stark choice: either migrate to the United States or deal with dwindling opportunities on the island. Raquel Salas Rivera captures this uncertainty and frustration in her poem "sinverguenza with no nation"; evoking the rage of Allen Ginsburg's epic poem "Howl," she tells us, "i saw the best souls of my generation /swallowed by colonialism."

While current residents feel increasingly pushed towards exile, the government has focused on attracting new "stakeholders" to come to Puerto Rico under Act 20/22, a pivotal piece of legislation that allows wealthy elites from the United States to use Puerto Rico as a tax haven. Passed in 2012, the legislation was created to bring capital investment to the island once it was barred from borrowing. Government officials promised that these newcomers, lured by seductive tax breaks, would invest in the local economy and create jobs. Under the statute, transplants from the United States who spend half the year on the island can receive exemptions from federal and local taxes, capital gains tax, and taxes on passive income until the year 2035, regardless of whether they generate employment or invest in the local economy. This makes Puerto Rico the only place under US jurisdiction where such income can go untaxed.[11] Of course, this is available only to "new arrivals," not current residents or those originally born in the territory who migrated and might wish to return. The trickle-down logic of Act 20/22, however, failed to make any positive impact on the lives of the majority of Puerto Ricans and only fueled the growth of hyper-segregated elite foreign enclaves around the island.

The economic crisis had thus already set the stage for what Puerto Ricans could and would come to expect after María. The public services and infrastructure that failed with deadly results during and after Hurricane María were already severely weakened after being deprived of the funds necessary to perform even minimal maintenance, let alone desperately needed upgrades. While public infrastructure continued to deteriorate, these forms of state abandonment normalized the idea of individual responsibility in the face of state retreat. As a result, in the days after María—while the local government was MIA, the power grid collapsed, and communication systems all failed—local residents fended for themselves. As Ana Portnoy Brimmer puts it in her poem "If a Tree Falls in an Island: The Metaphysics of Colonialism," there was a feeling that while Puerto Ricans struggled to make themselves heard, "only the ocean responds/with a swallow."

Refusing to be swallowed up and disappeared by inaction and silence, Puerto Ricans continued to shout their truth and take recovery into their own hands. Families, neighbors, coworkers, congregations, and groups that were already working together on social issues, as well as others who weren't, began coming together into self-described "brigades" to clear out roads, bring food and water to the forgotten and the vulnerable, distribute tarps, and eventually build roofs and homes. All the while state aid, both local and federal, refused to arrive.

Activists' efforts in the wake of María drew from and amplified grassroots organizing that predated the storm. This organizing was already focused on supporting individuals and communities in the face of prolonged economic and social crises. The notion that only the people would and could save the people—that the state could not meaningfully improve the lives of Puerto Ricans under the current political and economic structure—was already guiding the work of many activists and organizations, as poignantly demonstrated in the contributions by Giovanni Roberto, Arturo Massol, Mari Mari Narváez, and Sarah Molinari. These grassroots efforts have taken on a new urgency and necessity after Hurricane María as Puerto Ricans are forced to deal with not only the physical destruction caused by the storm but also the further destabilization of many communities as the government failed to act or acted in ways that only exacerbated vulnerability.

Two years after the storm, local residents continue to mend their society. They have come together to deal with the long-term issues of the slow recovery—lack of public lighting, unrepaired infrastructure, denials of FEMA assistance and insurance claims, and the loss of thousands of jobs—that have forced many to reimagine their lives, often beyond the geographic bounds of the Puerto Rican territory. Meanwhile, the local government claims it is open for business and continues to lure new arrivals with the promise of tax incentives, government contracts from emergency funds, and the assurance of a "resilient" population that can continue to adapt to the challenges that lie ahead. While the local government is working to sell post-María Puerto Rico as a blank slate onto which millionaire investors can project their wildest fantasies of unrestricted capital growth, Puerto Ricans on the island and in the diaspora are drawing from a rich history of resistance to construct a Puerto Rico for Puerto Ricans and deal with the storm's continuing aftershocks.

FROM DISASTER TO DECOLONIAL FUTURES

Building from the premise that Hurricane María is not a singular event, the contributors to this volume document the many shocks that Puerto Ricans have endured before and after the storm. Through reportage, poetry, personal narrative, and scholarly investigation, they show that the effects of Hurricane María are best understood as the product of a long-standing colonial disaster. The events unfolding in Puerto Rico appear to be following the familiar scripts of "disaster capitalism," for-profit recovery, and economic austerity that have been deployed in other parts of the world. Yet, as Yarimar Bonilla and Naomi Klein discuss in the dialogue that opens the volume, these policies take on a particular hue in the context of Puerto Rico. As Klein argues, in most disaster-struck societies, moments of crisis are viewed as opportunities for social and economic engineering. In the case of Puerto Rico, however, there was no need for elites to scramble to take advantage of new opportunities because there was already an infrastructure of dispossession and displacement firmly in place. Because of the financial crisis, a Fiscal Control Board had already been instituted by Congress to expedite economic transformation. Moreover, given Puerto Rico's colonial history, the board easily wielded neoliberal technologies—such as tax loopholes and financial incentives for foreign capital along with the fast-tracked privatization of local resources—characteristic of an economic system geared toward extracting wealth and catering to foreigners. For Klein and Bonilla, this is why since María the government has shown such little interest in creating new social and economic models for Puerto Rico, turning instead to the well-trodden paths of extractive Caribbean economies, with mere cosmetic updates for the digital age.

It is partly for this reason that Nelson Maldonado-Torres closes the book by asking us to consider whether María is truly a crisis, which he describes as a moment that requires a critical decision, or a disaster: a moment when decisions seem to have already been made and fates tragically cast. Maldonado-Torres is not the only one to point to the deep historical roots of María's disaster and the role of colonialism in defining its outcomes.

Several of the contributors in the volume describe what can be called the *coloniality of disaster*, that is, the way the structures and enduring legacies of colonialism set the stage for María's impact and its aftermath. For example, Frances Negrón-Muntaner examines how a liberal rhetoric of inclusion in which Puerto Ricans are framed as "fellow Americans" sought to make Puerto Rican suffering legible and worthy of moral outrage by the American public. Yet these accounts ultimately entrench the divisions between Puerto Ricans as a minoritized group and those "real" Americans called upon to care. Similarly, Hilda Lloréns shows how deep-seated racio-colonial tropes shape the way Puerto Ricans are depicted as "climate refugees" in the midst of a massive exodus from the island. Lloréns argues that these depictions play on the long-standing idea of Puerto Rico as part of the "disastrous tropics," an idea that functions not only to occlude the role of US colonialism in shaping and creating disasters but also to position migration to the United States as a salvation for Puerto Ricans rather than another form of disastrous uncertainty. Both Chris Gregory and Erika Rodríguez discuss how local photojournalists tried to push back against these representations by using different photographic techniques, refusing to reproduce the traditional visual scripts of victimhood in disaster reportage. Rodríguez, who was working on a photo series exploring how the colonial relationship between the United States and Puerto Rico has shaped Puerto Rican identity, addresses how this thread remained in her work while covering the storm. She says she felt a responsibility to document the dignity of Puerto Ricans "beyond the loudness of the disaster," in order to combat precisely the kinds of historical tropes that Lloréns traces. Similarly, Gregory's use of portraiture, his compositions, and his choice of subjects contest representations of Puerto Ricans as mere hapless victims of nature. At the same time, the palpable sorrow evident in his representations—conveyed through not just portraiture but also landscapes and still lifes—hints at the complex emotional fabric of post-María Puerto Rico.

As many contributors demonstrate, the particular trauma experienced in Puerto Rico after María is deeply tied to a longer preexisting colonial trauma. Several of the contributors point to how María's impact on the local psyche pushes us to think more critically about questions of dependency, self-sufficiency, sustainability, and sovereignty. Both Benjamín Torres Gotay and Eduardo Lalo suggest that Puerto Ricans' feelings of abandonment and the trauma experienced in the face of governmental and imperial neglect are rooted so deeply in colonial logics that they become "unnameable," as Lalo puts it. Torres Gotay describes how local residents seemed almost incapable of recognizing and assimilating the abandonment they experienced. In the face of FEMA's denials and the state's retreat, residents offered up narratives of resignation, repeatedly asserting that things "could be worse" and that, all things considered, they were OK. Torres Gotay roots this dissonance in the contradictory imperial narratives that Puerto Ricans have been fed about their place within a US empire.

Recognizing the long-standing and historically informed nature of trauma in different Puerto Rican communities, activists and scholars such as Patricia Noboa stress the importance of providing not only economic and legal assistance but also psychotherapeutic accompaniment. Noboa offers a compelling ethnographic description of how marginalized communities, long accustomed to state abandonment, deal with losses that extend far beyond the material. She shows that while some mental health professionals focus solely on the specific trauma of María, the storm triggered connections to past traumas that remain unmentioned and unrecognized for many. Noboa also challenges the discourse of "Puerto Rico se levanta" (Puerto Rico will rise) that emerged after the storm. She notes that such rhetoric pathologizes the difficulties of overcoming trauma as a result of individual shortcomings and a lack of resilience while failing to hold the state accountable for the structural violence that predated the storm.

Many contributors point to the difficulty of narrating this trauma. As Lalo suggests, some writers can only point toward what cannot be put into words. We see this absence of language throughout the texts. For example, Beatriz Llenín Figueroa's contribution is a Hurricane diary that never was, a chronicle that remains unwritten. Sofía Gallisá offers a series of lists that seek to take inventory of what was lost; as fragmented shards of testimony, they cannot be fully sutured into a narrative whole. In her contribution, Carla Minet discusses how local journalists dealt with the lack of transparency and accountability, particularly in regards to the death count, which for many remains an unforgivable act of government deception. While the dedicated work of journalists, lawyers, and activists has helped us start to grasp the scope of the storm's human impact, Minet reminds us that a total accounting of the storm's effects remains painfully elusive. Noboa suggests that even in the face of these unnameable experiences, it is important to listen, to witness, and to create spaces of narration. Rather than promoting a frenzied rush toward "recovery" without assessing what was experienced, what was lost, and what was transformed, these contributors encourage us to dwell in fractured narratives that emerged from Hurricane María and its aftermath, suggesting that even in their incompleteness these fragmented tales reveal powerful, if difficult, truths.

In the face of these silences and dissimulations, the arts take on a role of central importance in helping us understand the effects of the storm and its aftershocks. The play *¡Ay María!*, which was performed around the island in the immediate aftermath of the hurricane, was produced by a group of independent actors as a way of coping with their own personal experiences of the storm. It showcases the full range of human experiences that characterized everyday life during the storm, from the poignant to the absurd. Marianne Ramírez and Carlos Rivera Santana also provide chronicles of artistic efforts that emerged after María both in the island and the diaspora. Rivera Santana details how visual art has become an important site of catharsis as Puerto Ricans confront the effects of both natural and man-made disaster. In her essay, Ramírez discusses how visual artists have been engaging with the *longue durée* of colonialism in Puerto Rico and using their art to assert a decolonial aesthetic and cultural sovereignty. Meanwhile, Richard Santiago offers a firsthand account of the pain and difficulty of becoming an artist in exile. Santiago also shows the importance of the arts in revealing ugly and painful truths that the government seeks to keep invisible, such as the death and human devastation that occurred in the storm's aftermath. Lastly, Adrian Roman describes his experience as an artist from the diaspora traveling to his family's hometown in order to gather and preserve the discarded pieces of broken lives. These artworks once again point to a collective experience that cannot be fully expressed in words, much less in a political program.

Multiple contributors show that María is not just about economic exploitation and social inequalities, but also about a deepening crisis of imagination. In the face of immediate matters of life and death it can be difficult to think beyond the current political binds toward new collective possibilities. In her essay, historian Mónica Jiménez encourages us to look to the past in order to rediscover revolutionary impulses that can help us think about solutions to the problems confronting contemporary Puerto Rico. Jiménez examines how the Puerto Rican Nationalist Party and its leader, Pedro Albizu Campos, when confronted with economic instability and natural disaster during the early twentieth century, warned against relying on the United States to provide solutions. Instead, they argued that Puerto Ricans needed to assert an economic and political sovereignty in the face of colonial immiserization. Indeed, we see in Jiménez's contribution as well as others' in this collection that the past is prologue not only to the current crisis but also to the kinds of radical political thinking needed to build new futures.

Throughout the volume, contributors demand something more than a mere recovery, if by recovery we are to understand a return to a previous state of affairs. In a number of pieces we see bold calls to break with the reigning social, political, and economic structures that produce disasters and that continue to rock Puerto Rican society. In the contributions by Ed Morales, Natasha Brannana, and Eva Prado we see how local activists are pushing to reimagine the ties of obligation and debt that bind Puerto Rico to the United States through a critical interrogation of the fiscal crisis. Sarah Molinari describes how residents came together to feed their communities, clean up their surroundings, and lend each other support in the face of bureaucratic violence. Sandra Rodríguez Cotto narrates how in the face of both a governmental and telecommunications collapse, a small community radio station was able to provide comfort and community to those who were left alone in the dark. Mari Mari Narváez calls on Puerto Ricans to apply relentless pressure on both the local and federal governments to act in more accountable and transparent ways. She argues that only by centering the needs and desires of Puerto Ricans will Puerto Rico be able to meaningfully function as a free and democratic society. Both Arturo Massol and Giovanni Roberto examine how the search for new social relations in Puerto Rico is about not just self-sufficiency but also moving from mutual support toward new forms of collective self-determination. This shift requires healing the many traumas and shocks that Puerto Ricans have faced, including the displacement and dispossession felt by the millions of Puerto Ricans who have found themselves, by choice or circumstance, forging their lives beyond the geographic confines of the Puerto Rican archipelago.

Overall, the contributors to this volume ask us to consider what it would truly mean for Puerto Rico to recover from the devastation of Hurricane María. Is recovery simply measured by a return to the conditions that marked life in Puerto Rico before the storm? If so, that would mean a return to the status quo of extraction and exploitation necessary for colonial capitalism to function. The essays that follow suggest that this would represent not a recovery but simply the continuation of a colonial disaster. Ultimately, Hurricane María forces us to reckon with not only the disastrous effects of climate change, particularly on already vulnerable people, but also the need for decolonization to serve as the centerpiece of a just recovery for Puerto Rico and the Caribbean as a whole.

1. Andrea González-Ramírez, "Life, Interrupted: The Invisible Scars Hurricane María Has Left on the Women of Puerto Rico," *Refinery 29*, March 19, 2018, https://www.refinery29.com/en-us/2018/03/193315/women-mental-health-puerto-rico-hurricane-maria.
2. Andrea González-Ramírez, "Life, Interrupted."
3. Departamento de Salud, "Estadísticas Preliminares de Casos de Suicidio Puerto Rico, Enero—Diciembre, 2017," http://www.salud.gov.pr /Estadisticas-Registros-y-Publicaciones/Estadisticas%20Suicidio /Diciembre%202017.pdf and https://www.usatoday.com/story/news /2018/03/23/mental-health-crisis-puerto-rico-hurricane-María /447144002/.
4. Andrea González-Ramírez, "After Hurricane María, A Hidden Crisis of Violence against Women in Puerto Rico," September 19, 2018, https://www .refinery29.com/en-us/2018/09/210051/domestic-violence-puerto-rico-hurricane-maria-effects-anniversary; Claire Tighe and Lauren Gurley, "Official Reports of Violence Against Women in Puerto Rico Unreliable After Hurricane María," Centro de Periodismo Investigavo, May 7, 2018, http:// periodismoinvestigativo.com/2018/05/official-reports-of-violence -against-women-in-puerto-rico-unreliable-after-hurricane-maria/.
5. Jeremy Scahill, "Hurricane Colonialism: The Economic, Political, and Environmental War on Puerto Rico," *Intercepted*, podcast audio, September 19, 2018, https://theintercept.com/2018/09/19/hurricane-colonialism-the -economic-political-and-environmental-war-on-puerto-rico/.
6. US Department of Treasury, "Puerto Rico's Economic and Fiscal Crisis," https://www.treasury.gov/connect/blog/Documents/Puerto_Ricos_fiscal_challenges.pdf.
7. Mary Williams Walsh, "Puerto Rico's Governor Warns of Fiscal 'Death Spiral,'" *New York Times*, October 14, 2016, https://www.nytimes.com /2016/10/15/business/dealbook/puerto-rico-financial-oversight-board.html; EFE, "Pasan ley de Emergencia Fiscal," *El Nuevo Herald*, April 6, 2016, https://www.elnuevoherald.com/noticias/finanzas/article70337262.html
8. Luis J. Valentin Ortiz, "Una Pueblo Quebrado, una Quiebra Costosa," Centro de Periodismo Investigativo, November 1, 2018, http:// periodismoinvestigativo.com/2018/11/un-pueblo-quebrado-una -quiebra-costosa/.
9. Shereen Marisol Meraji, "Puerto Rico's Other Storm," *CodeSwitch*, September 19, 2018, https://www.npr.org/templates/transcript/transcript.php?storyId=649228215.
10. Larry Elliot, "Austerity Policies Do More Harm than Good, IMF Study Concludes," *Guardian*, May 27, 2016, https://www.theguardian.com/business/2016 /may/27/austerity-policies-do-more-harm-than-good-imf-study-concludes.
11. Jesse Barron, "How Puerto Rico Became the Newest Tax Haven for the Super Rich," *GQ*, September 18, 2018, https://www.gq.com/story/how -puerto-rico-became-tax-haven-for-super-rich.

PART I

Openings

THE TRAUMA DOCTRINE

A Conversation between
Yarimar Bonilla and Naomi Klein

YARIMAR BONILLA. I wanted to start by asking you to speak about your decision to write directly about Puerto Rico. You've written indirectly, and perhaps not even knowingly, about Puerto Rico for so many years. Your work has been so important for thinking about what's happening in Puerto Rico, not just in terms of María but also in terms of the debt crisis and the various political and economic transformations taking place there. What I particularly want to hear is your perspective as a non–Puerto Rican and as someone who has thought about these issues on a global scale. What is the particularity of Puerto Rico for thinking about the broader issues of disaster capitalism and the shock doctrine? What role does colonialism play in giving these relationships a particular hue?

NAOMI KLEIN. When I was in Puerto Rico researching this book, we also made this short film that you referenced.[1] I worked with a wonderful cinematographer named Christian Carretero. He often referred to María as "our teacher." He said, "María was a very strict teacher—we learned a lot about how we're living. The weaknesses of that, the vulnerabilities of that." Many painful lessons that I know we're going to be talking about. I just want to acknowledge that many of my teachers are in this room. The people who gave me a crash course in colonialism as it has impacted Puerto Rico in these five hundred years. You [Yarimar] are one of these key teachers, and you are quoted in the book and the film. I begin the book with Arturo Massol, with this beacon that is Casa Pueblo—this pink house lit up because they had solar panels on their roof and had another model for community-controlled and community-owned renewable energy. Eva Prado, who leads the movement calling for an audit of Puerto Rico's debt, was so patient when I met her in Puerto Rico and also as I was writing the book in helping me understand the illegality and the odious nature of Puerto Rico's debt.

Your question about the particularity—every place is particular. Many of the places where I have studied this process that I call the "shock doctrine" or "disaster capitalism" have a very painful history of colonialism and slavery. We have these intersecting forces . . . I think that the place that the particular intersections and layering of crises remind me the most of when I think about Puerto Rico, is New Orleans after Hurricane Katrina. People were talking about a "natural disaster," but of course there was nothing natural about it. You had this storm that was supercharged by climate change slamming into infrastructure that was deliberately neglected. You remember the warnings, "Repair the levees, repair the levees." And the levees weren't repaired because the people on the other side of those levees were considered disposable. They were the poorest people in New Orleans. Then you had the layering of white supremacy on top of that; you turn on Fox News and there was this animalization of the victims who had been abandoned in the Superdome. All these lies that were perpetuated in right-wing media about "they're raping babies," and so on. I think there were a lot of parallels.

One of the things that made it different was what I call in *The Battle for Paradise* "shock after shock doctrine." I'll read a quote that I use in the *Shock Doctrine*. It's a useful one from Milton Friedman, the architect of neoliberal economics in many ways. He wrote this in 1980: "Only a crisis, actual or perceived, produces real change. When that crisis occurs, the actions that are taken depend on the ideas lying around. That, I believe, is our basic function. To develop alternatives to existing policies, to keep them alive and available until the politically impossible becomes politically inevitable."

Natalie Jaresko, the executive director of the Fiscal Control Board in Puerto Rico, was interviewed by NPR recently, and the interviewer asked her, as somebody coming from Eastern Europe, what lessons she brought to Puerto Rico, and she said, "The lessons I think I bring to this are to use the moment of crisis—this fiscal crisis, this hurricane crisis. Use the political will in the moment to do the most that you can to change the structure of the economy."[2] That's a pretty clear articulation of the shock doctrine. In these moments you usually have a scramble to come up with those ideas, but in Puerto Rico the ideas were all ready to go, because of the preexisting debt crisis.

For instance, with Katrina, there was an emergency meeting at the Heritage Foundation two weeks after the levees broke. It was chaired by Mike Pence, who was then the head of the Republican study group that was the right-wing caucus on Capitol Hill. It had all the right-wing think tanks, like the American Enterprise Institute. They came up with a wish list. The waters had not yet receded in New Orleans, and they came up with a list: don't open the public schools, give parents vouchers that they can use in private schools, support the creation of charter schools, close the public housing projects, have a tax-free free enterprise zone. You go down the list, and there's thirty-five of what they call "free market solutions" to Hurricane Katrina and high gas prices—they just tacked on the high gas prices, by the way. It's an unbelievable list, because on this list, they say, "Open up the Arctic Wildlife Refuge to oil drilling" . . . what the hell is that doing on the list? What's amazing about it is that you have this crisis that comes from the intersection of climate change—this superstorm—and a deliberately weakened and neglected public sphere. And the solution is to do away with the public sphere and supercharge climate change by digging up more oil. Just do everything you can to make it worse.

One of the things that's different with Puerto Rico is that that infrastructure of crisis exploitation didn't need to be scrambled together and there didn't need to be a brainstorming session, because the Fiscal Control Board, locally referred to as "la Junta," was already in place. They already had all the policies; they didn't need to do any more planning. All they needed was the bloody-minded opportunism to push it through. Not harnessing political will, that's a lie; it's harnessing trauma, the state of emergency, the fact that people are just struggling to stay alive, and using that dislocation to ram through a preexisting, totally articulated agenda. This is why we heard about the push to privatize electricity after Irma, before María even made landfall.

You asked specifically about colonialism. One of the most powerful things I heard when I was researching this book was from somebody named Juan Rosario, who some of you know is a longtime environmental activist and labor activist in Puerto Rico. He talked about how colonialism is a war on the imagination, and how in these moments when you have opportunistic players coming in with their ideas for their "Puertopia" and plans for further privatization and deregulation, the legacy of colonialism has made it very difficult for Puerto Ricans themselves to come in and say, "No, this is our plan." But when I was there, I found that there was more confidence than in any place I've ever studied, to come together with a plan—with a people's plan. And you [Yarimar] mentioned Junte Gente,[3] and that's why I decided to do the book, so we could get money to this coalition—not nearly enough, but some. It is so remarkable; whether New Orleans after Katrina or Iraq after the invasion, I've certainly seen a remarkable ability of people under extraordinary circumstances to come together and say no to the shock doctrine, to disaster capitalism. But what I've never seen before is what I saw in Mariana, which was a meeting in a community that still didn't have electricity, except for the solar panels they put up themselves, that still didn't have water, that was still feeding itself in a community kitchen.[4] Where the schools were still closed. For people to come together under those dire circumstances and try to come up with a political program. It hasn't come together as fast as some people would like, but it is still extraordinary, absolutely extraordinary, that people were able to engage in that level of forward-looking organizing in the midst of an ongoing disaster. And that has to do with, I think, the infrastructure of resistance to colonialism and to the Junta.

I want to throw it to you, Yarimar, because you know so much more about this than me. And when we met in San Juan and I interviewed you for this, what struck me most was that you were in a really unique position to document and theorize this layering of multiple shocks because you were already engaged in this research before María hit. You were able to go back to some of the people with whom you were already discussing how the economic crisis and the debt crisis was being exploited. You were able to go back to them and ask, "So now what is María doing to your life?" Your book is yet to be published—it's going to be incredible—I'm very excited about it. But, I would love for you to share some of what you've learned in the early stages in those conversations that were punctuated and interrupted by María.

YARIMAR BONILLA. I had been doing research in Puerto Rico since 2015, and originally, my project was about the statehood movement in Puerto Rico. I was interested in doing an ethnographic study of those who imagine that making Puerto Rico the fifty-first state of the United States is the best option for decolonizing Puerto Rico. It's a political movement that has not been seriously studied, and in particular, it has not been analyzed as an anticolonial option. I was also interested in using the statehood movement to turn a mirror onto the United States, to open up the question of why the United States acquired territories that were from their origins not destined for full inclusion—in other words, how and why the United States became a colonial empire that refuses to recognize itself as such.

As I was working on that project, the fiscal crisis, which had been unfolding, was publicly announced. The governor declared that Puerto Rico was in a debt spiral, the PROMESA law was passed, and the fiscal board was appointed—the antidemocratic Junta that was now deciding the fate of Puerto Rico. Moreover, a series of Supreme Court cases in 2016 made Puerto Rico's colonial status and the limits of its sovereignty more apparent. At that time, there were dovetailing discussions about the economic crisis and "the death of the ELA," the Estado Libre Asociado, or commonwealth status. For many, the economic crisis represented the end of the mirage of prosperity that the ELA represented. The recognition by the Supreme Court that Puerto Ricans did not have self-government, not even the ability to declare themselves bankrupt, was the final nail on the coffin. There were even performative funeral rites that were held with demonstrators symbolically carrying the ELA in a casket. And spontaneously many artists and activists began creating murals, T-shirts, and placards featuring the Puerto Rican flag painted in black. For some this represented a symbol of mourning, while for others it was a sign of a new era of resistance.

I thus became interested in thinking about the current moment as one of "political death." My focus thus expanded beyond the statehood movement to think more broadly about the economic crisis in Puerto Rico as also a moment of political crisis. Indeed, part of the problem with colonialism is that in Puerto Rico, it's locked us into a set of options none of which fully represent what Puerto Ricans really want, but which at the same time limit our ability to think beyond them. As Christine Nieves Rodriguez, an activist from the Barrio Mariana in Puerto Rico, once stated: "Puerto Rico's greatest crisis right now is its crisis of imagination."

I left Puerto Rico three days before Irma, and I arrived in New York with a writing fellowship and an outline of the book I was going to write about Puerto Rico's political and economic crisis. Then Irma hit, and then María hit. Of course, given the work I had been doing, I was primed to think about the hurricanes in relationship to the political and economic transformations already underway in Puerto Rico. My focus since has thus not been on what María has caused, but what it has revealed: centuries of colonialism, decades of economic crisis, and deep forms of structural and infrastructural neglect.

It's important to keep in mind that these issues were already being challenged before the storms. In 2017 Puerto Rico had the largest May 1 demonstrations in years, and there were already many new forms of activism building against austerity, against the debt, against the Junta. María reshuffled a lot of those efforts, but people were able to move quickly and to engage in important political and social actions immediately after the storm precisely because of that already-existing political momentum.

It's partly because of this that I argue that what we see in Puerto Rico is not a shock doctrine but a trauma doctrine: this is not a case of economic and political interests taking advantage of a moment of shock, but rather of corporate and political interests taking advantage of deep-seated colonial traumas that have left the

population vulnerable to exploitation, all too accustomed to abandonment and self-reliance. The much-touted resilience of Puerto
Ricans thus needs to be itself understood as a form of trauma: years
of abandonment by local and federal governments have forced
communities to take care of themselves. I think this is why people
were able to move so quickly and to immediately begin thinking
about alternatives at the community level. This is wonderful but
also troubling, given the superhuman capacity for resilience that is
now expected of residents.

That's part of why I wanted to start the conference today with
the play *¡Ay María!* to open up our emotions, to dislodge what many
of us here have been holding in. I sometimes worry that everyone
moved so quickly to deal with what was happening that they didn't
take time to process the near-death collective experience that they
had suffered and the trauma generated from the large-scale structural
abandonment. The ability of residents to move through these experiences without taking a moment to mourn what happened—mourn
their dead, mourn their losses, and mourn the harm that they've
experienced—is not necessarily a healthy form of resilience.

The traumatized resilience that Irma and María compounded
and illuminated brings us back to the question of sovereignty in
Puerto Rico. In this moment many are reimagining what sovereignty could be and what it might mean. But I believe the question
we need to ask is: What is the relationship between self-reliance and
self-determination? On the one hand, you have communities taking responsibility for their own future: thinking about how to feed
themselves, how to guarantee access to drinking water, how to construct sustainable forms of energy, and so on . . . Yet, on the other
hand, you have the state pushing forward austerity measures, gutting public services, shrinking the education system, while they also
welcome what they call "new stakeholders" to the island.

This includes some of the folks that you talk about in your book:
the cryptos, or the Bitcoiners, or the "Puertopians," as they were
imagining themselves at one point. I think the extent to which they
are going to become major players is still an open question. The governor has said that blockchain is going to lead the recovery, but I'm

not sure *he* even knows what that means. What I do know is that these wealthy new arrivals are increasingly gaining power, becoming a political constituency—lobbying officials, writing op-eds in the newspaper, gaining political traction. Since María, the governor has been touring the United States—from New York to Silicon Valley—advertising post-María Puerto Rico as good for business. He presents Puerto Rico's colonial status—its lower wages, tax loopholes, and regulation exceptions, and now even the resilience of the population—as business opportunities. He calls Puerto Rico the ideal place for "the human cloud": untethered workers who can set up with a laptop anywhere. Again, I'm not sure if he knows what this means, but I'm interested in how these cryptocurrency entrepreneurs themselves are thinking about sovereignty, and why it is convenient for them to be in a non-sovereign state. A lot of them are libertarian ideologues who don't believe in a strong state to begin with, and so they see Puerto Rico's non-sovereign status as something they can really take advantage of. Of course, others in Puerto Rico could really use public services, labor legislation, environmental protections, and other signs of a stronger state. So, it remains to be seen how these dueling interests will become reconciled.

Your book takes up many of these questions. Can you speak a little more about how you see this unfolding?

NAOMI KLEIN. There's a clear overlap between the Bitcoin entrepreneurs who have descended on Puerto Rico and this obscure movement that some of you may have heard about called seasteading. Some of the players are the same. Seasteading is a movement that started about a decade ago. It was funded by Peter Thiel, a billionaire, a James Bond villain figure who shuts down media outlets who write things he doesn't like. He's an extreme libertarian, devotee of Ayn Rand. He funded an organization whose president is Milton Friedman's grandson. And they have this dream of building floating city-states that are completely sovereign, where they get a literal blank slate because they build these floating islands. They've been talking about it and theorizing for a decade now, and they've confronted some technical problems around setting up in international waters. They imagine them to be green, they imagine them to be totally self-sufficient. There are all kinds of prototypes on it. Another interesting thing about Peter Thiel as it relates to climate change is that he's bought a huge amount of land in New Zealand, where he's building his postapocalyptic bunker, which was featured in the *New Yorker* in an article called "Doomsday Prep for the Ultra-rich." This is important for us to think about in the context of climate change: why it seems feasible to deny the reality of climate change in public while privately building your own bunker. We're seeing a lot more of that.

There is this overlap, and there are some people who have been involved in the seasteader movement who are also active in the cryptocurrency economy. Their dream is to be free from government, free from taxes, to have total sovereignty—to have their own society. This is their liberation movement. They see any form of taxation and regulation as an attack on their freedom. There are all these logistical problems in building their sovereign floating city-states, and then along comes Puerto Rico and these laws that offer pretty much corporate Club Med: 4 percent corporate tax, no tax on dividends, no tax on interest, no capital gains tax. It's very appealing, particularly for the Bitcoin crowd, because they want to cash out and convert their cryptocurrency into real currency, and they don't want to get taxed on it. There are huge amounts of money at stake here. I think we need to understand this in the context of the depopulation of the island, where part of the draw is that Puerto Rico doesn't get totally depopulated, but a lot of Puerto Ricans leave Puerto Rico; they get to build their city-states, which is one of the things they're openly talking about.

So, you have that vision of sovereignty that I've just described: a very thin idea of sovereignty, where sovereignty means hyperindividualism; it means "I'm accountable to no one." This is in contrast to the vision of deep sovereignty that we're hearing more about, which is not only political sovereignty but also energy sovereignty, food sovereignty, and water sovereignty. This second kind of sovereignty is all about interdependence within and between communities and with the natural world. This is why I call the book *The Battle for Paradise*—

because we have these two dueling visions of utopia that could not be more different. One precludes the other. They can't happily coexist because the land is scarce and the need for a tax base is so important.

But there is a shift happening. I noticed this when I was talking to younger organizers in Puerto Rico—people who have grown up since the global economic crisis in 2008 and who had watched the way that supposedly sovereign states like Greece had their sovereignty stripped from them. Yanis Varoufakis, the former finance minister of Greece, says that "governments used to be overthrown with tanks; now it's with banks." I think there's more of an awareness for the generation that has come of age post-2008, and witnesses this stripping of sovereignty, and also with the memory of Katrina, and how places like Detroit and Flint have had emergency managers imposed on them, that sovereignty is not only about achieving political independence. They are not abandoning the project for political sovereignty but are really trying to understand what it means in the age of global capitalism, when you can have international financial institutions and national governments strip the sovereignty of states using all kinds of financial levers. Strip the sovereignty of cities, strip the sovereignty of whole nation-states, or you just have sovereignty on paper, but you just have no economic control. That is what has happened to Greece.

This is not to say political sovereignty doesn't matter, but I think there is a growing interest in the idea of "multiple sovereignties," a phrase I first heard from Barcelona's mayor, Ada Colau, in the context of the fight for Catalan independence. She supports the Catalan people's right to determine their relationship with Spain. But unlike more traditional sovereigntists, she has continued to insist that sovereign borders are not enough—we need housing sovereignty, we need energy sovereignty, we need water sovereignty, we need deep sovereignty. And we have to protect the space to talk about those multiple sovereignties because there is a way that fights that revolve around natural borders can tend to erase all these other sites of struggle and just say, "We'll deal with all of that afterwards. First, we have to get political sovereignty." And then often it comes at the expense of all that.

Before we move on, I think you need to share a little bit about your research with the crypto entrepreneurs, because you've been interviewing them. What have you found in talking to them about what their vision is for the island? What do they want?

YARIMAR BONILLA. Part of the problem is that they don't have a vision for the island; they have a vision for themselves on the island. What they want is a place in which to enact individual, not collective, sovereignty.

I have asked some of them why have they chosen to come to Puerto Rico, and they're very straightforward: for the tax incentives, and the nice weather helps. I've also interviewed some who have come after María, and I've asked them, "Why do you want to do business here? There's unreliable electricity, unreliable internet." And they have said, "Well, in Condado, the internet is better than in Silicon Valley, and the electricity never goes out." So, they have this kind of bunker mentality and a confidence that they have the personal resources to fill in for any lack in collective services. It is concerning that at a time when the government is reducing public services and enforcing austerity measures that disproportionately impact the poor, these new constituents could help tip the scales toward a recovery that centers on individual responsibility at the same time that local populations are left without collective safety nets.

They've also said to me, "After María, this was the worst thing that Puerto Ricans could possibly experience, and so if this is what the worst looks like, then we're OK." It's important to know that after María, hedge fund managers and investors were really paying attention to what was going to happen in Puerto Rico, and the value of Puerto Rican bonds actually went up because people were able to deal with the storm so well.

NAOMI KLEIN. Since I wrote *Battle for Paradise*, the repression of social movements and protest is becoming more and more severe. And there was a moment when the governor was just shaking with anger about the fact that projectiles were thrown during a May Day protest led by schoolteachers this year. The governor was so angry that somebody had thrown a rock. He said, "This is the message: we're trying to get investors to come here." So, what you're saying is so literal, he [the governor] is saying we need a subdued population that does not resist, because resistance is a direct threat to this vision.

YARIMAR BONILLA. Absolutely. The 2017 May 1 demonstration before María was one of the largest public demonstrations in recent history, while these most recent May 1 events became one of the largest instances of state repression. The police teargassed not just protesters but also bystanders and journalists. Not only this, but police went into surrounding communities and teargassed surrounding streets as well.

I think a lot of people in the United States who don't understand the history of economic exploitation and political repression in Puerto Rico would be surprised that the only things protesters are throwing at the governor right now are water bottles and pebbles. A lot of US onlookers expected rioting and looting after the storm. But there wasn't much, really. There were very few protests and very little violence. Instead, communities focused on taking care of themselves.

This is what's really complicated: how resilience can serve as a pressure valve for the state. And I think this is one of the big questions we need to tackle: How do we turn our self-reliance into something that can be self-determining rather than something that simply lets the state off the hook? Resilience suggests that business can carry on as usual, that Puerto Rico can be "open for business" even as thousands remain roofless, homeless, displaced, and destitute. In this way, resilience becomes jargon for simply adapting to, rather than confronting or transforming, unacceptable conditions.

NAOMI KLEIN. It reminds me of when I was in New Orleans after the BP disaster when the huge oil spill happened. I was talking to a civil rights lawyer named Tracie Washington. She was remembering that after Katrina, there was so much talk of the resilience of the people of New Orleans, and it wasn't interrogated too closely at that point. But then people started using the same discourse after the spill, which had a huge economic impact on the fisheries. And Tracie said, "I never want to hear the word *resilience* ever again. Resilience means you can hit me again." And I agree. But when I think about the work that Arturo Massol is doing with solar microgrids, it's important to state that we do know that decentralized renewable energy is more resilient and that there are more shocks to come. That's what climate change means.

YARIMAR BONILLA. I think we want our infrastructure to be resilient. We want our buildings, our electricity systems to withstand shocks, but we don't want our population to be required to withstand repetitive shocks and traumas.

And this should be explicitly connected to similar conversations going in mainland United States. For instance, African Americans in the United States are expected to withstand shocks to the point that they are believed to have higher thresholds of pain. So, when they go to the hospital, they don't receive the same kind of pain medication. This is directly connected to the notion that Puerto Ricans don't need the same wages, they don't need the same public services, they don't need the same access to basic services. How to manage the effects of these expectations while also pushing against them is one of the big questions.

What I think Massol and other activists in Puerto Rico are looking for is not resilience but sustainability—the ability to continue to exist in the long term, and to not just survive but to thrive. We cannot thrive under our current energy model or our current political model.

NAOMI KLEIN. One of the things that every "shocked" society—or anybody who has experienced a massive traumatic event—has in common is the feeling of a total lack of control over their lives and an inability to protect their loved ones. That is the most horrifying feeling that any individual can feel. Especially for the parents in the room, that feeling of "I can't project my kids"—it's the most wrenching feeling. And we need to remember that when thinking about the infrastructure of disaster capitalism, where Puerto Ricans are excluded from their own recovery and are put into this passive position of watching outsiders come in. Whether it's NGOs or whoever it is, it's actually retraumatizing, because once again the impacted people are out of control, and they are being acted upon, this time in the so-called recovery. They have no control.

And it's actually the opposite of recovery. Because the way to recover from that trauma of losing all control over one's life is to have some control once again. To be empowered to exercise control. That is healing. Everyone who works in trauma recovery knows this: that the way to help is to give people agency again so that they are participants, not spectators in their own lives. It doesn't erase the trauma, but there is a healing that happens.

And I saw some amazing examples of this in Puerto Rico. For example, one of the most moving experiences I had was at this farm school in Orocovis, where I met Dalma Cartagena, who has run the farm school there for eighteen years. The students learn agroecology as part of their education. What struck me most about visiting the school was that Dalma, unlike so much of the humanitarian industrial complex, immediately gave the students agency after María. She said, "You can help feed your families; you can be part of the healing process. Just by planting these crops and growing this food." The students were harvesting crops when we visited and were so energized by that work. They were so happy. The other thing Dalma talked about was how important it was for young people to learn to trust the natural world again. Because the feeling of being in a superstorm like María is to feel that the natural world has turned against you. Learning to trust the natural world, as a source of strength and sustenance, reminds us that we are part of a web of life, that the land can support life, and that we're part of this unending relationship. She told me, "I tell the students: touch the flowers, touch the plants, rebuild that trust." That kind of knowledge has to inform what recovery means. It can be woven in. People are doing that, but it's in these small pockets. The people with resources aren't learning from it at all, aren't interested in it at all. They are retraumatizing people by treating them as helpless when they aren't.

YARIMAR BONILLA. I think that's such an important point because as a new hurricane season approached in Puerto Rico a lot of people discussed how they get scared when it rains, scared when a stiff wind blows. There is so much basic trauma that hasn't been attended to. I think this idea of agency is so important, and so is the idea of a just recovery. Not a people-powered recovery in the sense of people being left to their own defenses to recover, but rather a recovery that offers justice for those who have experienced these repetitive shocks. Part of the problem is the way that Puerto Rico has been discussed in the media: it has become a Trump story. Puerto Rico would have gotten much less coverage if Trump hadn't been in office. But the media was not really focused on how Puerto Ricans were imagining their recovery, and on how Puerto Rico fits within a larger Caribbean story. They were only interested in finding Trump's Katrina. In addition, the insistence that we must pay attention to Puerto Ricans because they're United States citizens is troubling. It's not because they're US citizens that we should pay attention, it's because they're experiencing a humanitarian crisis. Creating this distinction between citizens and noncitizens in the Caribbean, and between citizens and noncitizens in the United States is really problematic at a moment when noncitizens are under attack. Moreover, it shifts the discourse away from a model of recovery that takes into account colonial histories and the search for regional solutions.

NAOMI KLEIN. I think it would be amazing if Puerto Rico could host a conference on how disaster capitalism is impacting the Caribbean right now, because it's all over the place. Barbuda is an extreme example. For those of you who don't know, and it's received almost no coverage, Irma led to a total evacuation of Barbuda. There was absolutely no one on Barbuda because of the evacuation. Barbudans were evacuated to Antigua, and you have this very unequal relationship between the government of Barbuda and Antigua, where Barbuda is a tiny player in the government of Antigua. The Antiguan government can make fateful decisions for the people of Barbuda, so it's kind of like a subimperial relationship. Barbuda before the hurricane had this extraordinary land law that makes it illegal to buy and sell land, not just to foreigners, but land is held communally in Barbuda, the legacy of a slave uprising. In a way, it's one of the very few examples of real land reform after slavery. That was hard won; it was a result of an uprising, and long protected, fiercely protected. Prime Minister Gaston Brown went right in and said this is our chance, this is our opportunity, Barbuda is open for business. Immediately, when the island was evacuated, he moved to change the land law. The other thing I would add about the problem with making this all about Trump and all about the Republicans is that stories like the ones in Barbuda don't get covered. Part of the reason they don't get covered is that, so far as I can tell, one of the single biggest beneficiaries of that changed land law in Barbuda is Robert De Niro, hero of the #Resistance here in the United States. He owns a large hotel in Barbuda and has been trying to get around that land law for years. I cowrote a piece about this on *The Intercept*, but the pickup on this was very limited because people were busy celebrating Robert De Niro for saying "Fuck Trump." I think that is just one vivid example of the limits of this frame and the problems with making it all about pathologizing Trump.

YARIMAR BONILLA. Yes. And this is the problem with pretending that Puerto Rico's problems begin and end with Trump, because they don't. I think it's important to keep this in mind, especially as it relates to the campaigns to get those who migrated after María registered to vote. First, it's important to note that these new arrivals are displaced populations. They are refugees of climate change and of a political and economic crisis. They shouldn't bear the burden of voting out a government that they didn't get to vote into power in the first place. Second, we shouldn't assume that they're going to feel part of a political process that they've been excluded from, disenfranchised from, their entire lives. To suddenly place these burdens on them turns them once again into colonial pawns. I don't think it's the responsibility of María refugees to flip states, or to flip Trump anything other than the bird.

1. Lauren Feeney, "The Battle for Paradise: Naomi Klein Reports from Puerto Rico," *Intercept*, March 20, 2018, http://theintercept.com/2018/03/20 /puerto-rico-hurricane-maria-recovery.

2. Michel Martin, "Hurricane María's Devastation of Puerto Rico, 1 Year Later," *All Things Considered*, NPR, September 23, 2018, https://www.npr .org/2018/09/23/650956637/hurricane-marias-devastation-of-puerto -rico-1-year-later

3. "JunteGente," accessed February 14, 2019, http://juntegente.org/en/

4. "Arecma," accessed February 14, 2019, https://www.apoyomutuomaría na.com/eng.

¡AY MARÍA!

Mariana Carbonell, Marisa Gómez Cuevas,
José Luis Gutiérrez, José Eugenio Hernández, Mickey
Negrón, Maritza Pérez Otero,
and Bryan Villarini

¡Ay Maria! actors in front of the residencial Plaza Apartments en Manatí
November 13, 2017. Image provided by Mariana Carbonell

EDITORS' INTRODUCTION

A month after Hurricane María ravaged Puerto Rico, a small group
of actors came together in San Juan to create a short play, *¡Ay María!*

(Oh María!), about their experiences before, during, and after the crisis. In a time when most of the archipelago still had no electricity or water, people still lived in shelters, and telecommunications were spotty at best, this group embarked on a mission to relieve some of the population's anguish and trauma through entertainment. During only a week of workshops, they collectively wrote the script and set out to present *¡Ay María!* in all seventy-eight towns of Puerto Rico in a rented RV. Sometimes the performances took place where there was a group of people who had lost everything they owned, as they waited for a free meal, a ham-and-cheese sandwich and a bottle of water, the only meal they would have that day. Sometimes the audience would join in the action, blurring the lines between fact and fiction. For the five weeks that it took the group to travel the archipelago, the actors looked directly into the sadness in the eyes of their audience and transformed that pain into a smile, into hope.

PRODUCER'S INTRODUCTION BY MARIANA CARBONELL

October 20, 2017, was supposed to be the opening night of the first play I produced in Puerto Rico. Instead, I found myself among friends from the San Juan theater community, commiserating over the damage the theaters had suffered after Hurricane María. It was exactly one month after the storm, and we were worried about when we'd be able to get back to work. I remember thinking of Peter Brook's words: "I can take any empty space and call it a bare stage." One thought led to another. If an empty space in San Juan is a stage, what about Mayagüez? What about Fajardo? "Whatever happened to traveling theaters?" I asked. This was the beginning of the idea that would become *¡Ay María!*

I reached out to acclaimed theater director and friend Maritza Pérez Otero, who practices collective creation in a style based on the Theater of the Oppressed. I told her I wanted to create a show about the hurricane and present it all over Puerto Rico, using theater to relieve some of the population's anguish and trauma through entertainment. I wanted this to happen as soon as possible. She enthusiastically agreed.

Recruiting the actors was harder. I must have reached out to more than twenty actors. Some could not join because they were working with FEMA, some never got my messages because we still didn't have reliable cellular service, some had to take care of elderly family members. In the end, five actors showed up for our first meeting, and what was supposed to be a brainstorming session became something more like group therapy. One actor had lost every physical possession he had, another lost her car to the floods. We all had a story to tell, for the hurricane had spared no one.

After only a week of workshops, together we developed the text of the show under Maritza's direction. The work centered around our experiences before, during, and after the crisis. The tragicomic vignettes and songs depicted the struggles shared by everyone. The lack of supplies, including water, gasoline, and batteries, the floods, the humiliation of Trump's visit, the failures of the local and federal agencies, and the constant battle with mosquitoes.

¡Ay María! actors at the CREARTE parking long in Yabucoa on November 26, 2017. Image provided by Mariana Carbonell

One of the most important aspects of the production was that it needed to be self-sufficient. I didn't want us to go to a town, perform, and then ask for a place to sleep, or a working bathroom, or an electrical outlet, or food and water. This is why the recreational vehicle, which we named Rocinante, was essential to the production. I rented the RV for a month. We performed the thirty-minute play two to four times a day in order to reach our goal of seventy-eight towns. None of the theaters in Puerto Rico were operating at this time, and some towns don't have theaters, so the venues for the performances varied widely. We would try to establish communication with the towns' mayors, school principals, or community leaders to coordinate a time and place for the performances, but oftentimes, especially in the island's central mountain range, communication was impossible, so we improvised. For five weeks we performed in schools, shelters, town squares, nursing homes, basketball courts, bakeries, and bars.

¡Ay María! actors at Parque Ciudad Masso in San Lorenzo on December 6, 2017. Image provided by Mariana Carbonell

As I look back on the experience nearly a year and a half after the hurricane, I get very emotional. I get sad because I remember the stories I heard, the destruction I saw. I get angry because I saw the conditions people were living in two months after the storm. The hurricane was a disaster; the government's response was a tragedy. I get frustrated because some of the themes explored in the play are as relevant today as they were in the days immediately after the storm. How we will ultimately recover from this is yet to be seen.

OH MARÍA!

Translated by Carina del Valle Schorske

¡Ay María! actors at the Ricardo Rodríguez Torres public school in Florida on November 13, 2017. Image provided by Mariana Carbonell

MICKEY. Hello. My name is Mickey, but now I'm a palo de guayaba [guava tree].

JOSÉ EUGENIO. Hello. My name is José Eugenio, and I am a palo de roble [oak tree].

BRYAN. Hello. My name is Bryan, and I am a palo de ceiba [ceiba tree].

MARISA. Hello. My name is Marisa, and I am a palito de ron [shot of rum].[1]

JOSÉ LUIS. Hello. My name is José Luis, and I am a hummingbird. (*Flies between branches.*)

JOSÉ LUIS. Crisis! A great disaster draws near.

MICKEY. I know what it is! It's the Fiscal Control Board.

MARISA. I know, it's the governor.

JOSÉ EUGENIO. I know, it's the ashes of Peñuelas. (*Everyone sneezes.*)[2]

BRYAN. No, it's another strike at the university.

EVERYONE, *chanting*: Candela, candela, la iupi da candela.

JOSÉ LUIS. No! It's a category 4 hurricane.

EVERYONE. Uy! (*Singing.*)
Marullo grande de mi amor
Creciente encantadora
Tsunami exquisito de pasión
Tu amor me envuelve como ola, ay qué ola.

[My love's great swell

Growing, enchanting
Passion's sweet tsunami
Your love overcomes me like a wave, oh what a wave! (*Repeated.*)]

MARISA. Hello and good morning. My name is Ada Bombón.[3] It's 6:35 in the morning, and the cone of uncertainty has become certain catastrophe.

MICKEY, *on his cell phone*. Mami, did you listen to Ada? She says the storm's coming in the form of a cone. I'm scared shitless. Put gas in both cars.

MARISA (ADA BOMBÓN). We're registering sustained winds of 275 miles per hour. According to the National Hurricane Center, Hurricane María is headed for Puerto Rico from the east/northeast.

BRYAN, *on his cell phone in the aisle of an imaginary grocery store*. There aren't even hot dogs? I'll buy beer. God will provide the rest.

MARISA (ADA BOMBÓN). This storm is among the biggest and strongest seen this hurricane season. I urge everyone to prepare themselves, to take all necessary measures.

JOSÉ EUGENIO, *on his cell phone*. There's no water in the supermarket. Or the pharmacy. What? The ATMs aren't working?!

MARISA (ADA BOMBÓN). Buy canned food, secure your storm windows, do everything you can before the gusts of strong rain arrive.

JOSÉ LUIS, *on his cell phone*. I'm in the battery aisle. There aren't any AAs, AAAs, or AEE left.[4]

MARISA (ADA BOMBÓN). Finish all your preparations. Leave the flood zones. Get out of vulnerable buildings. I'm urging you, please: stay calm!

(*Everyone screams and runs. Bryan spins like the hurricane with a mosquito net over his head. While he lists the names of Puerto Rico's municipalities, all the characters start falling, one on top of the other.*)

BRYAN. Naguabo, Aibonito, Utuado, Utuado otra vez, Arecibo. (*He takes off the mosquito netting.*) I'm dizzy, I'm going to Tampa. (*The pileup of people takes notice.*) José Luis! Come up here, it's flooding down there! (*José Luis stands up.*)

BRYAN. Marisa! And your mother? (*Marisa stands up.*)

MARISA. I think she's in the shelter. Mickey! (*Everyone helps Mickey stand up.*) How many times did I tell you to go to the shelter?!

MICKEY. Girl, don't give me a hard time.

JOSÉ LUIS. Are you OK?

MICKEY. I lost everything. (*Pause. Singing.*)
Fue un 20 de septiembre
Cuando nos cogió María
No me había llegado el agua
Desde la dichosa Irma.
Yo me fui pa casa'e mami
Yo me fui pa casa'e papi
Yo me quedé en mi casa
con mi gata bien paría . . .

[It was the 20th of September
When María came for us
We said we were blessed
When Irma's waters spared us.
I went to my mami's house
I went to my papi's
I stayed in my house
where my cat was giving birth]

(*Individually, members of the cast ask members of the audience where they spent the night of the hurricane.*)

MICKEY. We owe this to . . .

EVERYONE, *singing.*
María bonita, María del alma
Que provocaste muchos derrumbes
E inundaciones y estos mosquitos

[Beautiful María, María of my soul
Thanks to you we have landslides
And floods and mosquitoes]

(*The cast huddles under the mosquito netting.*)

JOSÉ EUGENIO. You wouldn't happen to have any more of the OFF that you lent me?

JOSÉ LUIS. What's this about OFF? What I have is a natural recipe for repellant from my grandmother and great-grandmother and great-great grandmother!

JOSÉ EUGENIO. Fine then, give me that recipe!

JOSÉ LUIS. Take out a paper and pencil and write it down. Start with a cup of extra virgin olive oil.

JOSÉ EUGENIO. Can it be vegetable oil?

JOSÉ LUIS. It could even be gasoline, if that's what you want.

JOSÉ EUGENIO. Virgin vegetable oil.

JOSÉ LUIS. Add a cup of wonder water.[5]

JOSÉ EUGENIO. Is that like the sweat of Wonder Woman?

JOSÉ LUIS. It smells better.

JOSÉ EUGENIO. Wonder water.

JOSÉ LUIS. Then you're going to add a few drops of eucalyptus oil.

JOSÉ EUGENIO. Could it be malagueta, sweet pepper?

JOSÉ LUIS. It could be malagueta, but grind it up good.

JOSÉ EUGENIO. Ground sweet pepper.

JOSÉ LUIS. Then add a few slices of cinnamon bark.

JOSÉ EUGENIO. Like the sticks?

JOSÉ LUIS. Like the sticks. Whole.

JOSÉ EUGENIO. How many?

JOSÉ LUIS. However many you want.

JOSÉ EUGENIO. Four sticks.

JOSÉ LUIS. Finish it up by adding cloves, what we use for making majarete and tembleque. Shake it up well. Then grease up your whole body with it, and there you go. An old remedy against mosquitoes.

JOSÉ EUGENIO. Neighbor, forgive me for questioning you, but this must smell awful . . .

JOSÉ LUIS. Not at all. This smells better than any colonia.[6]

MARISA. Well, sure, boys, anything's better than this colony.

José Eugenio. Where did you find all those things?

José Luis. You have to wait in line at the store.

José Eugenio. Uyyy, that line is really long.

(*The cast emerges from the mosquito net and gets in line. While they talk they slump lower and lower until they collapse on the floor.*)

Marisa. This line isn't as bad as the one for water.

José Eugenio. True, it's not as bad as the line for the ATM.

José Luis. Yeah, it's actually not as bad as the line for buying a little bag of ice.

Mickey. At least this line's moving a little faster than the one for gasoline.

Bryan. This one's not as bad as the one for Western Union.

Marisa. I was waiting longer to get into the supermarket.

José Eugenio. This one's worse than the line for the emergency room.

José Luis. I was in line for a week to get a power inverter.

Mickey. I even had to wait in the line for diesel.

Bryan. Excuse me, do they distribute tarps here? (*Singing.*)

Fue un 20 de septiembre	[It was the 20th of September
Cuando nos cogió María	When María came for us
Trump llegó a Puerto Rico	Trump arrived in Puerto Rico
Después de 13 días.	After thirteen days.

Y el gobernador le dijo	And the governor said,
"aquí no ha pasado nada"	"Nothing's happened here"
Y el tirano anaranjado	And the orange tyrant
Nos tiró papel toalla.	Threw paper towels at us.]

MICKEY (TRUMP). Hello, brown people of Puerto Rico. I'm the president of the world. Your governor told me there have been only sixteen deaths. That's not a disaster. The real disaster is the way you're messing with my budget!

MARÍA (SAN JUAN MAYOR YULÍN). Our people are dying!

MICKEY (TRUMP). Shut up, nasty woman. Here, I've brought you extra soft toilet paper to wipe your little butts.

JOSÉ LUIS (ROSSELLÓ). The plan for the planned planification is based on an extensive planity plan that I planned all by myself planning in the . . .

(*While Rosselló talks, Bryan translates his speech into sign language at his side. Mickey, Marisa, and José Eugenio play football with a roll of paper towels till they hit Bryan in the face with it and he falls down "dead."*)

MICKEY. And Mr. Governor, what about this fatality?

JOSÉ LUIS (ROSSELLÓ). Well, did they do an autopsy?

MICKEY. He just died!

JOSÉ LUIS (ROSSELLÓ). Well, it doesn't count.

JOSÉ EUGENIO. Mr. Governor, my best friend committed suicide a week after the hurricane because he'd lost everything. Does that death count?

JOSÉ LUIS (ROSSELLÓ). Well, did they do an autopsy?

JOSÉ EUGENIO. The morgues were all full!

JOSÉ LUIS (ROSSELLÓ). I see, well, then it doesn't count.

MARISA. Mr. Governor, what about the hundreds of people who couldn't receive medical care without electricity, which was caused by the hurricane. Do they count?

JOSÉ LUIS (ROSSELLÓ). Did they get autopsies?

MARISA. The administration ordered the cremation of hundreds of cadavers without autopsies!

JOSÉ LUIS (ROSSELLÓ). Who approved this?

MARISA. Héctor Pesquera.[7]

JOSÉ LUIS (ROSSELLÓ). Ah well, they don't count.

MICKEY. Mr. Governor, and the baby from Barranquitas who was carried away by the wind right in front of her parents?

JOSÉ LUIS (ROSSELLÓ). Well, was there an autopsy?

EVERYONE. The wind ripped the baby away!

JOSÉ LUIS (ROSSELLÓ). Well, it doesn't count. The press conference is over. (*He answers his cell phone.*) Daddy, did I do a good job?[8]

BRYAN (REVIVED). Is there cell service?

(*Everyone looks for service on their phones.*)

BRYAN. Titi Luli? Titi Luli? You sent me a generator from New York? Oh no—by mail? They've already stolen it.

(*Everyone gets in a line. Mickey and José Luis are carrying a baby in their arms.*)

JOSÉ EUGENIO. Welcome to the office of the public advocate. Please form a line. How can I help you?

JOSÉ LUIS. We're still waiting on an electronic deposit to our Family Link card.

JOSÉ EUGENIO. I'm sorry, love, if there isn't electricity, how are you going to receive an electronic deposit? I can't do anything for you.

MICKEY. You can't do anything for us?

JOSÉ EUGENIO. Do you know the convention center in San Juan? That's where the COE is. They can help you there.[9]

MICKEY. But the COE told us they can't help us!

JOSÉ EUGENIO. Next! How can I help you?

MARISA, *talking fast.* Look, the bank took my house and the hurricane swept away the house that I'm renting so I left for a shelter but we're too many families in this school building and the wastewater started rising so we left for another shelter but this one isn't in my town and I don't have a car so I don't have any way to take my kids to school, I need a house, I can't go on sharing a bathroom with forty-five people and when you see the empty houses that the bank has taken from people, all these foreclosed houses with roofs, totally empty . . . listen, I'm desperate, I don't know what to do . . .

JOSÉ EUGENIO. Breathe! Look, next to the convention center is the Sheraton hotel. That's where the governor is with all his friends enjoying the air-conditioning . . . go complain there. Next. (*Bryan steps up, and José Eugenio pinches his nose.*) What a stench!

BRYAN. So, this is why I'm here. In my community there's no water, and we want to know when the trucks of water will arrive, those ones they're calling "oasis" . . .

JOSÉ EUGENIO. I recommend going to Roosevelt Roads in Ceiba because there are a bunch of water bottles over there covered in tarps.[10] You shouldn't drink them, but at least you can bathe yourself with them, because with that stink I can't do anything for you. (*Mickey draws close to José Eugenio and almost vomits on him.*) Uy, leptospirosis, get away from me, go to a hospital. Next.

JOSÉ LUIS. I applied for $500 from FEMA, and it still hasn't arrived.

JOSÉ EUGENIO. Do you see the sign over there? That's the number for FEMA. Everyone use that for your calls, because I'm on a shortened schedule. Goodbye.

MARISA, *at a FEMA call center.* You have reached the Federal Emergency Management Agency. To continue in English, say "English."

(*Everyone at once.*)

BRYAN. English.

MICKEY. Español.

JOSÉ LUIS. Espanish.

JOSÉ EUGENIO. Spanish.

MARISA. To begin, press the five-digit zip code where the damage occurred.

(*Everyone says a different zip code.*)

MARISA. I'll transfer you to an operator. "Hello, FEMA."

MICKEY. Hello. Am I talking to a human being?

MARISA. Mmhmm.

MICKEY. Thank God, because I don't understand the machine. Look, I'm calling because I applied for the $500 they're giving people . . .

MARISA. Yes, the $500. Lots of people have called for the $500. Unfortunately, if it has not arrived in your account, you have to call this same number to appeal it.

MICKEY. But I've been on hold for three hours. Appeal it for me.

MARISA. I'm sorry, I don't have the authorization to appeal it . . .

MICKEY. Please!

MARISA. You're going to have to call again. I'm sorry.

MICKEY. Don't hang up!

MARISA. Have a good day. (*Hangs up. Answers.*) Hello, FEMA.

JOSÉ EUGENIO. Hello. Look, I have a problem, it's raining and I'm getting wet.

MARISA. Well, go inside your house!

JOSÉ EUGENIO. I am inside my house! But my roof is like a colander. I need a tarp.

MARISA. Yes, there are lots of people asking for tarps, but we don't have any left. The problem is that María came too late. We already had Hurricane Harvey in Texas and Hurricane Irma in Florida, and unfortunately, we just don't have enough tarps for all these emergencies. You'll have to go to Home Depot.

José Eugenio. But we pay for FEMA, and I don't have extra money to buy a tarp.

Marisa. I'm sorry. You can try calling back tomorrow. (*Hangs up. Answers.*) Hello, FEMA.

José Luis. Hey, baby. Is military food microwavable?

Marisa. How many times have I told you not to call me at work? Don't come at me with some bullshit about military food. You're a grown man, you can make rice. Finally someone in this home has a job, someone got a little gig with FEMA. I need you to take care of the kids. Don't call me at work. (*Hangs up. Answers.*) Hello, FEMA.

Bryan. Hello! FEMA? Yes. English! Because in English the money comes faster!

Marisa. Hello? (*Hangs up.*)

Bryan. Hello? Hello? (*Looks at his phone.*) My battery ran out!

(*Bryan and José Eugenio become a military helicopter. Marisa, Mickey, and José Luis yell at the soldier like little children excited to see a helicopter. Bryan throws out packets of food in slow motion.*)

Bryan. Military food. . . . Very high in sodium . . . It gives you crazy constipa . . .

(*The helicopter takes off, leaving children coughing in the dust.*)

Mickey. I don't want to be a soldier anymore!

(*Everyone becomes generators on the floor, making generator noises.*)

Bryan. There's a reason "Puerto Rico rises".[11] Who can sleep with this noise?!

MICKEY. And this smell.

MARISA. And this heat.

JOSÉ EUGENIO. Good morning, neighbors!

JOSÉ LUIS. Ave María, you're fresh today!

MICKEY. José, you're going to have to do something about this generator. It doesn't let me sleep. I just can't.

JOSÉ LUIS. You should set it on a timer so that it turns on at a certain hour.

MARISA. And I need you to do something with that muffler. My kids have asthma, and you know how the hospitals are in this country. I can't take that risk.

JOSÉ EUGENIO. You complain now, but not when I pass you the extension cord to charge your stuff.

MICKEY. Look, José Eugenio, the question is, Is the beer cold?

JOSÉ EUGENIO. Since last night!

(*Everyone celebrates.*)

JOSÉ LUIS. Well, then leave the generator on!

MARISA. In last night's domino game we decided whoever lost would have to cook tonight.

JOSÉ LUIS. Bryan!

BRYAN. Me? I can't even boil water without burning it! The kitchen isn't for me.

José Luis. OK, well, I'll do Bryan a favor and cook.

José Eugenio. If you want, I have some frozen steaks that we can share among the neighbors.

Mickey. Ay, I'm going to miss you.

José Eugenio. You're leaving too?

Mickey. It's not that I want to leave, it's that I have to. The house doesn't pay for itself; the light, the water, and the telephone have been cut off for two months, and they're still charging me. And the only call coming in is my boss telling me that the business is closed and I have to look for another job. Neighbor, I have savings, and I've tried to stretch that money, but it's running out. So, I have a cousin that left for the States and is living in some tents for refugees in a parking garage, and he tells me that it's fine and that I should come with my daughter. Because that's the other thing: my daughter has a chronic condition and she used to get therapy in school. But my daughter's school hasn't opened yet, first because it was a shelter and then they closed it like so many schools that Keleher is shutting down.[12] And my daughter's teachers are responsible, decent people, so they went to protest the closing of the school, and can you believe they were arrested?! Try and tell me this country isn't fucked. Neighbors, this country is strangling me. This country with its corrupt officials is messing with my family and yours. What do you want me to do?

Marisa. I can lend you money.

Mickey. And how will I repay you?

José Eugenio. You're talking nonsense! What's this about going to live in a tent city in a parking garage? You have neighbors here who support you and will lend a helping hand when you need it. The sun is going to come out again.

MICKEY, *singing to the tune of "Sale el Sol" by Ismael Rivera.*
¿Saldrá el sol
Dentro de esta tragedia?
Vivo desesperado
Pensando en el avión.

Todos menos Mickey
Estarás en otra tierra, no en la tuya
En un trabajo que no te gusta
Estarás en otra tierra sin tus hijos
En un invierno pasando frío

From this tragedy?
I live in desperation
Dreaming of airplanes.

Everyone but Mickey.
You'll be in another country, not yours
In a job you don't like
You'll be in another country without your children
Trudging through a cold winter]

MICKEY. But I have to leave.

MARISA. If you have to leave, then go, and join the diaspora that's helped us so much from abroad.

MICKEY, *to the audience.* Neighbor, what do you think? Should I stay or should I go?

(*Various answers from the audience.*)

MICKEY. OK, I'll give it two more months. But if I stay, I'm not going to stay quiet, because I'm angry. The only good thing about those storm winds? They ripped through and exposed the government's scams for all to see.

MARISA. Now we can see how badly administered this country is, from the Department of Consumer Affairs to the Department of Housing, Health, and Education. While people were going hungry, the government was signing multimillion-dollar contracts with American companies, and you get to thinking, what has the government ever done for us?

MICKEY. I know what they did, they made a hashtag: #PuertoRicoSeLevanta. But it's been a long time since Puerto Ricans and everyone who lives here, even if they aren't from here . . . it's been a long time since we've been on our feet. Here, the one who has to stand up is the government.

EVERYONE, *singing.*
    Marullo grande de mi amor
    Creciente encantadora
    Tsunami exquisito de pasión
    Tu amor me envuelve como ola, ay qué ola.

    Al otro día de María
    Al despertar los vecinos
    Con hacha, machete y pico
    Abrimos nuevos caminos, caminos.

    Y levantamos escombros
    Y compartimos comida
    Pero pa' buscar el agua
    hicimos tremenda fila, ay qué fila.

    Así queda demostrado
    Quien levantó nuestra tierra
    La gente trabajadora
    Que somos una jodienda, jodienda.

    Ay María
    Hay caminos

Los vecinos
Construímos
Puerto Rico

[My love's great swell
Growing, enchanting
Passion's sweet tsunami
Your love overcomes me like a wave, oh what a wave!

The day after María
All the neighbors woke up
With axes, machetes, and picks
We opened up new roads, new ways.

And we cleared rubble
And we shared food
But to find water
We waited in line, what a line it was!

Now it's clear
Who lifted up our country
The hardworking people
We're the shit!

Oh María
There are ways
we neighbors
rebuild Puerto Rico.]

1. In Spanish, the word *palo* can refer to both a tree and a shot or drink of liquor. The original Spanish text draws a parallel between the guava, oak, and ceiba trees and Marisa's little cup of rum, between the natural world and the world of human comforts.

2. Peñuelas is a town on the southern coast of the island where the private company Applied Energy Systems (AES) dumped toxic coal ash. Activists had been struggling against this pollution since before the hurricane. But now the problem is even more grave: the ash was not covered in preparation for the hurricane, and we don't yet know the effects of this negligence.

3. This is a reference to meteorologist Ada Monzón, who became infamous during María for her frequent Facebook Live messages.

4. AEE refers to the local power authority: Autoridad de Energía Eléctrica.

5. Agua Maravilla (wonder water) is a popular brand of witch hazel.

6. In Spanish, the word *colonia* means both colony and cologne.

7. Héctor Pesquera is the Security Secretary and became a controversial figure after he denied the death count.

8. Ricardo Rosselló's father, Pedro Rosselló, is a former governor of Puerto Rico.

9. The Centro de Operaciones de Emergencia (COE) was the controversial center of operations for the local government and FEMA after María.

10. Nearly a year after María, millions of spoiled water bottles were found to have been abandoned by FEMA on an open-air strip in this former military base.

11. "Puerto Rico se levanta" (Puerto Rico rises) was the post-María motto promoted by the government and private companies, taken up by many individual Puerto Ricans. "Se levanta" has many meanings—stands up, gets back on its feet, wakes up, rises—but all of them in this context emphasize self-sufficiency and bootstrap solutions rather than an ethic of state or even community care.

12. Julia Keleher is Puerto Rico's Secretary of Education and a polarizing figure because of her role in closing down hundreds of public schools in the name of austerity.

PART II

Narrating the Trauma

WAPA RADIO

Voices amid the Silence
and Desperation

Sandra D. Rodríguez Cotto

Imagine you are a thirty-year-old woman. You're a single mom with a daughter, and you have moved to Manatí, a town on the northern coast of Puerto Rico, in an economically deprived area. You are in hiding there from your estranged husband, who has repeatedly raped and beaten you. He once hit you so hard that you were left legally blind. You can see colors only with a special light that your daughter carries for you. You are afraid, shaking, fearing he might come back to kill both you and your daughter. She is six years old.

Now imagine that, in addition to all this, a hurricane knocks out the electricity for months. Roads are closed, debris blocks passage on the streets, and power lines are strewn on the ground.

This happened to a woman who, only God knows how, walked to four different police stations after the hurricane hit. The first one was destroyed by a river, and the other three had only one or two officers. She wanted to file a report stating that she was a victim of domestic violence and needed protection. She knew her attacker was following her, and she wanted to inform the community where she was living. The officers in the last station said they couldn't help because they had no cars and the power was out.

One night in San Juan, about three weeks after the hurricane, I was on the air at WAPA Radio, the only radio network that was covering the entire island, taking calls from listeners. We had six phone lines, but only one was working, so all the callers had to wait to get on it. People called asking for help or food, or to let their relatives know that they were alive. It was almost midnight when the woman called the station. It was difficult to hear her over the static. After all, the telecommunications system had collapsed, as well as radio and TV antennas. There was no internet either.

She said she needed help and was only asking someone to donate one of those special lights that would allow her to see. Hers had broken during the hurricane. She needed one to escape, she said, and her six-year-old daughter would serve as her guide. Hearing her voice, I immediately knew something was wrong, and I kept asking questions. Then she told me and all our listeners about her ordeal.

Just imagine her desperation—she was willing to talk on the radio, regardless of whether somebody identified her, in order to save her and her daughter's lives.

I told her to wait so I could take the phone call in private, but the call dropped just as she started to talk. I went crazy. I feared the worst. Frantically, I begged her to call back. I asked God to save her. This all happened on the air.

Luckily for the woman, a group of nuns in the town of Guánica, on the southern coast of the island, who belong to the Fátima Sisters Order, were listening. One of them knew the area because she has relatives nearby, and she made her group of nuns jump into their van and drive through the darkness, in the middle of the night, all the way north to find the woman. They did. The saved her and her daughter's lives and put them in a safe house near their convent. Two days after the call, I found out about the nuns and about a Pentecostal pastor and his wife who brought not one but two of the special lights to the radio station to donate them to her. The nuns came to the station and picked them up along with other donations.

My colleagues and I at the radio station heard stories like this from the day after the hurricane, September 21, all the way to December.

⁓&

Nobody is prepared to experience what we did at the station. Better said, nobody had any idea what the hurricane would lead to. Some of us remember Hurricane Hugo in 1989 and Hurricane George in 1998. Almost an entire generation had gone by without a hurricane. So initially, people would listen to the newscasts and the meteorologist, and they heard the messages telling them to be ready. But the truth is, most people thought nothing would happen.

Part of the public disbelief resulted from the local government's poor strategy and planning. The official government propaganda website, as I called it, belonged to the Emergency Management Office (Centro de Operaciones de Emergencia), but it was managed by the same public relations firm that ran the election campaign of Governor Ricardo Rosselló and that handles most official communications. Instead of sending out tweets about how to prepare for the hurricane, or even announcing a list of shelters, the firm tweeted modeling photos of First Lady Beatriz Rosselló in her seventh month of pregnancy. The same image appeared on several official Twitter and Facebook feeds, such as the police, 911, and the governor's office.

A few days before the hurricane hit, Rosselló did go on a media tour telling people to move to a safe location. But some people didn't take the warning seriously. This changed thanks to what I call "the Pesquera effect." The head of the Security and Public Safety Commission, Héctor Pesquera, went to the press and told the public, "You have to move or you are going to die." That's when people got scared. Pesquera was in charge of supervising emergency management, but his experience as a cop led him to threaten the population as a tactic to compel everyone to take the hurricane seriously. In retrospect, you can tell he had no idea how to handle an emergency, but his tactics scared enough people and made them move to safe locations. But it was too late.

Well before the hurricane, it was evident that there were problems in the preparation efforts.

A week before Hurricane Irma destroyed the Lesser Antilles and affected the east coast of Puerto Rico, I published an op-ed titled "The Dependency on Antennas" on the local news site NotiCel. I asked what would happen if a hurricane knocked out telecommunications. After it was published, I got a call from Sandra Torres, president of the Puerto Rico Telecommunications Regulatory Board. "That's not going to happen," she said. The telecom companies, she said, had spent millions in developing a solid infrastructure. I knew she was wrong. Companies were investing not in their infrastructure but rather in their strategies to increase their share of the market. I knew the technology in Puerto Rico was not up-to-date.

Then Hurricane Irma hit, and half the island was left without power and without telephone service. So, I wrote a second column in NotiCel titled "Telecommunications Collapse." And then María hit. It was complete chaos. People had expected interruptions in power and water service, but not telecommunications.

That really hit a collective nerve.

People were desperate. Entire families would park their cars on the highway to try to find a signal so that they could call their relatives and tell them they were alive. The younger generations had never experienced anything like this and were in a state of shock being without social media or the internet. It was like going back in time.

Without television, newspapers, or the internet, the only thing available was the radio. During the hurricane, I was listening to the number-one network, Univision Radio, on a tiny, battery-operated radio when renowned personality Rubén Sánchez said on air that the station's windows had shattered due to a fallen tree and they had to go. Then the station went off. It had lost its main antenna. Then Noti-Uno network went off, and so did Radio Isla. Those were some of the largest AM networks. The same happened to the FM stations. It was scary and unbelievable. After all, Puerto Rico has more radio stations per square mile—a total of 125 stations, including the second-oldest in Latin America—than any other US territory.

No one could believe everything was so silent. It was eerie as if death were near. I moved my tiny radio dial to WAPA Radio 680 AM, a mom-and-pop operation owned by a Cuban exile family whose motto is that Puerto Rico must become the fifty-first state; every day at noon they play "The Star-Spangled Banner." It is a small, very modest station, but the owners had spent thousands on improving their microwave and analog signal, and it was the only station that survived

María. The voices of their three hosts, Jesús Rodríguez-García, Luis Penchi, and Ismael Torres, were the only ones on the air throughout the entire island while the winds were destroying entire towns. WAPA Radio was the only network that remained on the air. There were over eight small stations operating in some areas, but their signals were limited to specific towns away from the hurricane's path. WAPA was the only network that could reach the entire island.

After thirty hours of uninterrupted broadcasting, the hosts' voices started to falter. Back home, I knew I had to do something. I needed to help. So, the day after the hurricane passed and winds were out, I checked that my house was OK, jumped into my car, and drove to the station. One of the owners, eighty-year-old doña Carmen Blanco, was surprised to see me and asked what I was doing there. "I just came here to help," I said. "Please go in and help," she replied.

I thought they would ask me to write the headlines, but they decided to put me on the air so they could take a break. I wasn't prepared, so I asked reporters to come to WAPA. And they did. Over sixty reporters from newspapers, television stations, online news sites, and even competing stations showed up at WAPA Radio to help. Many doctors, psychiatrists, and volunteers did the same and filled the station.

Hundreds of people formed long lines around the station every day, waiting to get in and write down their names so that the reporters would mention them on the air and let their relatives know they were alive. Some came asking for help, food, and water. Others had lost everything. Many churches, community groups, and nonprofits brought in donations. And then mayors, military personnel, and even Rosselló, came to the station because the government's emergency headquarters was flooded. For a few days after Hurricane María, WAPA Radio became the government's operations headquarters.

In that chaos, we aired twenty-four hours a day. To make this possible, the network's two other owners, Jorge and Wilfredo Blanco (Carmen's son and husband) had to maintain WAPA's network of six stations and their repeating antennas and microwave signals around the island. The stations are WMIA in Arecibo, WISO in Ponce, WTIL in Mayagüez, WVOZ in Aguadilla, WXRF in Guayama, and WAPA in

San Juan. So, they spent their days driving throughout the island along broken roads, from one antenna or station to the next, refilling gas generators and checking signals. They did that for four straight months.

We stayed at WAPA and continued our work as volunteer reporters. At night, we slept on the floor. We did this partly because people kept coming to the station hour after hour asking for help, partly because we were receiving so much terrible news of devastation, but especially because we are all journalists. We didn't leave the station because we had a strong ethical commitment to help people and report the news.

From day one, we asked hard questions that government officials couldn't or wouldn't answer. Why wasn't the head of FEMA providing information about the death toll, or ongoing environmental crises such as flooding? Why wasn't the secretary of health providing guidance to the public in order to prevent epidemics? Why was the secretary of Family Affairs (Departamento de la Familia) silent about the managers who abandoned nursing homes, leaving the elderly people there to die? Who could tell us how many people were left homeless? How was aid being distributed, and why did the mayors keep disputing the governor's claims that they were all receiving it? And most importantly, how many people had died? Eventually, the government simply stopped responding to us. The only answers they gave were to the reporters from US mainland media.

~&

I vividly remember what happened one of those long days in the town of Arecibo. People called WAPA to report that the town's mayor, Carlos Molina, had commandeered a power generator that FEMA had provided to a shelter full of over eighty elderly people. Molina had moved the elders—some of whom needed ventilators—into shelters without electricity in the nearby towns of Barceloneta and Camuy. The generator went to a restaurant called Arasibo Steakhouse, where, Molina claimed, it would be used to prepare meals for hundreds of rescuers. We went to Arecibo to investigate.

When we arrived at the Arasibo Steakhouse, we heard loud music and found what seemed like a party. It turned out that the mayor's employees were using the generator for a political fund-raising party. Ricky Martin's "Livin' la Vida Loca" was playing, and the employees were having a blast, while a few roads ahead people still had no food and some were dying in nearby communities. When we asked about the generator, a man became so upset that he almost hit one of the reporters, Francisco Quiñones. That man happened to be the president of the Municipal Assembly. Back at the station that afternoon, we aired the report, but nothing came of it. Two days later, Rosselló held a press conference in Arecibo and said he completely supported Molina, who is his political ally. Nobody investigated afterward.

That same attitude was on display for months from government officials in different municipalities and in the central government. The truth was not coming out. Without any news media, people were not informed. Many deaths went unacknowledged by top government officials.

Fifteen friends or relatives of mine died after Hurricane María. Fifteen. My very best friend, Aileen, was a forty-two-year-old woman, head of human resources for one of the island's largest supermarket chains. Day after day she had to deal with dozens of employees in need, some of whom had lost their homes. She had no power or water at home. The stressful environment took a toll on her, and she had a heart attack in her office. She died in front of several employees and her thirteen-year-old daughter, whose school was closed. Authorities sent her corpse to one of the infamous trailers outside the Forensic Sciences Institute and left it there for almost two months. The government claimed it had no personnel, and many families were in the same situation. I kept calling and calling different government officials, asking to have her body returned so we could have a proper funeral, but nothing happened. Her body was finally returned to her husband and parents on December 31. It was so badly decomposed, they had to cremate her immediately. I didn't get a chance to say goodbye to her. There was no funeral.

In the hurricane's aftermath, the government put the number of deaths at sixteen, then sixty-four; then a Harvard University study estimated that over 4,965 had died, and a study by George Washington University and the government put the death toll at about three thousand. Everyone knows there were many more. There are hundreds or maybe thousands of people who suffered, who became ill or whose illnesses were exacerbated by the lack of electricity, the impossibility of reaching a hospital in time because roads were destroyed and rivers overflowed, or other problems caused by the hurricane. These people will never be part of the official statistics.

People, not the storm, created much of the chaos. Puerto Rico's infrastructure had been abandoned for decades amid the island's recession and local partisan politics, and people died because the government failed to follow emergency plans established by previous administrations. Adding to all that was the racism and neglect of the Trump administration, which provided aid too slowly, and the corruption and ineptitude of the local government. This was a true recipe for a disaster.

One common denominator since the beginning of this emergency has been the insensitivity, cynicism, and arrogance of officials close to Rosselló. To maintain their message that only sixty-four people died from the storm, when so many Puerto Ricans were burying their relatives, was a mockery. It was disgusting to see elected officials and their relatives bragging on social media about how the local government was doing a "good job" or saying that because the dead do not vote, they are unimportant.

For all those reasons each of us reporters on the radio questioned the government's lack of transparency. Officials tried to hide the dead, but the stench of death prevented it. The dead were present, and their spirits demanded justice and respect. We have to honor the memory of our dead; we must tell their stories.

~❦

The collective pain touched almost everyone in one way or another, regardless of their class or location. It didn't matter if you lived in San Juan, up in the mountains, or in the diaspora. If you didn't lose a relative, you lost an acquaintance, had a friend who was sick, or knew about a family that was broken up as people fled the island. The pain was the same.

The people of Puerto Rico maintain a perpetual feeling of mourning when we think about what happened, about the dead, and about those left behind, even if government officials never admit the pain we live with. We can see it in how people act, in their absent gazes. We can see it in the sadness of those who are still struggling to recover and those who lost a loved one. These are the aftershocks of Hurricane María that no one can deny.

The hurricane and its aftermath are a political story, an economic story, and a colonial story, but for me, it was also a personal journey. It is a tragedy of losing many friends and relatives, the undeniable desperation of seeing—firsthand—people in need and the arrogance of politicians and government officials. For me, it is also a story about resilience and survival.

Looking back, I could say the aftershocks of Hurricane María changed me. Change, after all, is the law of life. I do not want to keep looking at the past, nor to think only about the future. We shouldn't forget what happened to us, but we must move on. We must be present in this moment.

Now I, like most people on the island, am more conscious of the need to prepare for disasters. But aside from saving canned food or storing water, one of the hurricane's better consequences is that we have realized we have to build strong neighborhoods and build communities. You have to learn to be friends, know your neighbors, and be able to help each other, as we all did across the island.

In that sense, the hurricane made me more open. It made me lose the fear of expressing my emotions in public, something that is hard to do as a reporter. But during these kinds of life-and-death experiences, objectivity does not exist. You assume the side of the truth and must defend those in need. When someone calls you live and tells you on the air, in the dark of night, that he is going to commit suicide, or cries from hunger, as so often happened to me working at WAPA, you learn to respond with your soul.

I will always carry my experiences on the air in my soul. I am the same as before the hurricane, yet I have also changed. I feel more alive and have a strong commitment to help create a better Puerto Rico.

Many months after the hurricane's first anniversary, I was talking to one of the nuns who saved the woman who was hiding from her abusive husband. The woman is now a regular worker at a women's shelter managed by the nuns. She is strong, resilient, and helping others overcome abuse. Her example taught me that we can overcome our fears and stand strong looking at the future. I know that's what we are doing in Puerto Rico.

Like our torn flag after the storm, we are resilient. Like our flag, still waving before the calmed sea, we must persevere.

MARÍA'S DEATH TOLL

On the Crucial Role of Puerto Rico's Investigative Journalists

Carla Minet

Two days after Hurricane María made landfall, Puerto Rico's Centro de Periodismo Investigativo (CPI), or Center for Investigative Journalism, regrouped. Everyone was so desperate to get to work that we all moved mountains to meet at El Telégrafo in Santurce, one of the very few places with Wi-Fi in San Juan. We started looking for a temporary newsroom, because we faced the same issues as most other citizens: no power, no water, no internet, no cell phone or landline service, no fuel, and no passable roads. There were almost no media outlets, and only one radio station was left standing. The government had collapsed, offering no official data and statistics.

After a week looking for places to work from, we ended up in the government's Center for Emergency Operations, which was the only place where we could find reliable internet, phone service, and power, as well as access to most government officials. We had to establish a whole new editorial agenda, which ended up focusing on two issues: the death toll and the hurricane's impact on Puerto Rico's colonial debt.

With Joel Cintron, Luis Valentín, and Omaya Sosa Pascual as reporters, our series about the debt reported on who owns it and what it meant in the new posthurricane context. Two months before the hurricane hit, Puerto Rico's Fiscal Control Board—imposed by the US government in 2016 to control the island's finances for at least five years—approved a fiscal plan setting forth how the island would repay its $74.7 billion debt and cover its $49 billion in pension obligations. The plan was based on assumptions about state expenditure that, in the hurricane's aftermath, were no longer true. Moreover, the plan did not take into account climate change risks, even though the island falls within one of the most vulnerable areas of the continent.

Our reporting became a point of reference, providing dozens of US and international news outlets the context to improve their stories and acknowledge Puerto Rico's systemic problems.

But the most important story we did was on the hurricane's death toll. It became a top story in Puerto Rican, US, and international news.

GOVERNMENT NUMBERS VS. GOOD SENSE

In the first seventy-two hours after the hurricane, Puerto Rican governor Ricardo Rosselló kept saying that, at most, sixteen people had died. At that point, our reporter Omaya Sosa Pascual had interviewed two doctors who together had seen nine deaths in one day. The official statistics made no sense.

Several days after the hurricane, the CPI began publishing reports revealing that dozens of additional confirmed deaths had not been reported to the government, because there was no special protocol to handle deaths during a disaster and because Puerto Rico's agencies responded slowly.[1]

Reporter Jeniffer Wiscovitch joined Omaya Sosa Pascual. For weeks, we interviewed doctors, police officers, rescue workers, funeral home directors, mayors and their staffs, and neighbors of the deceased. Based on their information, we built our own database. We also requested information from official sources, including the Health Department, the Security Department, the Police Department, the Demographic Registry, US Health and Human Services, and the Centers for Disease Control and Prevention. But they mostly had very little to say.

We started to wonder how the bodies of dead people in Puerto Rico were being handled, since there were no computer systems to register deaths and many morgues had no electricity to preserve the corpses. Our reporters collected reports and flyers about missing people and tips from the radio, social media, and community leaders. At this point, we did most of our reporting on the street in the most devastated areas; we knocked on doors and drove on dangerous and difficult roads.

One of the cases we reported was that of Teodoro Colón-Rodríguez, in Orocovis, a rural town in the center of the island. Colón-Rodríguez had died September 20. His corpse lay at home for three days before a funeral home was able to fetch him and agree to bury him without a death certificate. No government official responded to the family's death notice. He is still unaccounted for as a hurricane-related death in official records.

Dependent on oxygen, Colón-Rodríguez was recovering at home from a stroke he suffered the week before Hurricane María. He died in the middle of the storm at the home of his daughter and grand-daughter in Orocovis's Damián Abajo neighborhood. That same day, once the winds died down, his son-in-law Angel Luis Vázquez took a machete and courageously walked into town for help. After walking for four hours through landslides, a rising river, and thick brush, he arrived at the police station, where he was told that he could not move the body and that police could not recover the body because the roads were cut off. The town funeral home, Orocovis Memorial, also told him they could not get the body.

Vázquez had to return home on foot, use a generator to turn on the air-conditioning for the corpse, and move the whole family into the next room, because the home's second floor was completely destroyed. The next day nobody arrived. Desperate, Vázquez walked back into town and went to the Emergency Operations Center to plead for help. Although there was none of the required paperwork available to move a body, municipal first responders and the funeral staff agreed to help him. That same afternoon, a group of ten men tried to reach the body to remove it from the home, but the rains prevented them from reaching the residence, said rescue leader Willie Colón, who was also a relative of the deceased.

On the third day, the ten men returned at dawn and were able to remove the decomposing body by carrying it and dragging it among landslides, trees, and a flooding river so it could be buried. Neither the police nor a prosecutor ever showed up to certify the death. The funeral home confirmed that the Demographic Registry was not operating during those days.

A doctor also never came, even though Colón-Rodríguez had been a hospice patient at Aibonito's Mennonite Hospital. His son-in-law had to tell the funeral home that he was responsible for any problem that could arise from moving the body without following a legal process.

We found dozens of cases of like this one.

Even though our first databases were rudimentary and at times seemed useless, we kept digging and turned them into a real mine of stories that kept growing every day. It had become almost a quest to prove that the government was wrong.

After reporters Omaya Sosa Pascual and Jeniffer Wiscovitch published their series of stories about how the government was underreporting deaths, the official general data released finally confirmed that there were over one thousand more deaths in September and October than there had been in 2016.[2] In December 2017, when the official number of deaths was sixty-four, Governor Rosselló ordered a recount and thorough investigation.

In November, as part of that process, our reporters started a collaboration with Quartz (qz.com), disseminating an online form that collected hundreds of stories from Puerto Ricans who said their relatives died because of Hurricane María but were overlooked by the government.

In February 2018, the CPI sued the director of the Puerto Rico Demographic Registry, Wanda Llovet-Díaz, after multiple attempts to get her agency to offer information about post-María deaths. The lawsuit stated that it was a matter of "public information and of great public interest for the people of Puerto Rico."

Specifically, the CPI asked Llovet-Díaz for detailed information about the deaths recorded in Puerto Rico in 2017, in a complete database format, up until the most recent entry in the Demographic Registry's system; death certificates issued in Puerto Rico from September 18 to the present; the deaths registered, broken down by day and municipality; burial permits granted since September 18; cremation permits granted since September 18; and access to the notebooks in which permits granted to funeral and cremation homes are recorded manually in each of Puerto Rico's Demographic Registry offices.

Our team worked with the Inter American University Law School's Legal Clinic, which supports CPI litigation. CNN joined our lawsuit a few weeks later. In the end, we won access to the cause-of-death database, death certificates, cremation and burial permits, among other documents.

CROWDSOURCING AND COLLABORATIONS

The CPI did more than fifteen investigations or follow-up stories in eleven months, and we developed the website hurricanemariasdead.com with a database of 487 verified cases of people who died because of Hurricane María, following CDC protocol. Our stories were republished or quoted more than fifty times by national, US, and international media; we entered into partnerships with the *Miami Herald*, CNN, the Associated Press, NPR's *Latino USA*, and qz.com; and we won a lawsuit against the Puerto Rican government for official records. The study of the death toll ordered by the governor was done, but the recount was not. At the CPI, we continue to investigate with our independent databases.

On the eve of the hurricane's one-year anniversary, the CPI presented a collaboration with Quartz and the Associated Press in which the names of the dead were matched against government death records released by the Puerto Rican government in response to a lawsuit by the CPI. Together, we interviewed about three hundred families of the dead and reviewed the records of nearly two hundred others using the Centers for Disease Control and Prevention's criteria for certifying disaster-related deaths. The CPI led the project, which took more than three months to complete and involved dozens of volunteers, journalists, and experts.

Most of the cases in the project's database are considered indirect deaths, meaning they were caused not by winds or flooding but by the lack of power, drinking water, and medical supplies after the storm.

The project did not interview the patients' doctors, and the death certificates themselves make no link to María. The Puerto Rican government acknowledges that hundreds or thousands of deaths should have been classified as storm-related but weren't, because of doctors' lack of training on how to correctly fill out death certificates. Participation in this survey was voluntary; therefore, the sample was not representative of Puerto Rico's demography and was not used to extrapolate trends in causes of death and demographics.

The project analyzed mortality databases from the Puerto Rico Demographic Registry from 2014 to 2017 to calculate changes in demographics and cause-of-death rates across the whole population using standard grouping for fifty cause-of-death rankings from the International Statistical Classification of Diseases and Related Health Problems (ICD-10)—the global standard epidemiological diagnostic tool.

Adding to the death toll saga, CPI reporters worked on dozens of other stories, inevitably related to the emergency and the recovery process. Among other issues, these included the lack of legislation governing climate change preparedness, systemic problems with hospitals, the energy players coming down to Puerto Rico to get their share of the recovery pie, ongoing problems with forensic corpse management, austerity measures on top of the recovery process that affect the recovery itself, lack of official statistics updates such as domestic violence, and the high number of FEMA denials for individual assistance.

Naturally, after the emergency stage passed, we shifted our focus, and now the CPI team is concentrating on scrutinizing the recovery, which is one of our biggest challenges moving on. The death toll investigation is still central to many stories, especially those that try to explain why the healthcare system collapsed the way it did and who was responsible for it. It's also a reference that serves as a checklist for future preparedness. We keep on investigating of all that. More to come.

1. Omaya Sosa Pascual and Jeniffer Wiscovitch, "Dozens of Uncounted Deaths from Hurricane María Emerge in Puerto Rico," Centro de Periodismo Investigativo, November 16, 2017, http://periodismoinvestigativo.com/2017/11/dozens-of-uncounted-deaths-from-hurricane-maria-emerge-in-puerto-rico/; Omaya Sosa Pascual and Jeniffer Wiscovitch, "Delayed and without Resources: Puerto Rico's Police Did Little to Investigate Missing Persons after Hurricane María," Centro de Periodismo Investigativo, December 17, 2017, http://periodismoinvestigativo.com/2017/12/delayed-and-without-resources-puerto-ricos-police-did-little-to-investigate-missing-persons-after-hurricane-maria/.

2. Omaya Sosa Pascual, "Nearly 1,000 More People Died in Puerto Rico after Hurricane María," Centro de Periodismo Investigativo, December 17, 2017, http://periodismoinvestigativo.com/2017/12/nearly-1000-more-people-died-in-puerto-rico-after-hurricane-maria/.

(NOTE FOR A FRIEND WHO WANTS TO COMMIT SUICIDE AFTER THE HURRICANE)[1]

Raquel Salas Rivera

no one teaches us to accept death because death, that canned death,
stays empty inside: the great hole of fuck it that wants to devour
us. no one explains how we can become part of the impossible new
world that is tomorrow, or how we are supposed to avoid falling
into the perfect and permanent under eye circle we call facing the
day. mana, how not to understand? that is the question i avoid with
the organizational fervor of a rescue team that never arrives, but i'll
tell you this: desire isn't always followed by death. sometimes i run
into you in the street and you shine like an orb or a solar lamp, but
you are still worth more than all the generators (in case you haven't
been told a thousand times). y other times, without tilde, i. i. i. other
times, your words reach me like a fundraiser that explodes and tem-
poralizes truth, like an espachurrao (squashed? flattened? spread?)
aguacate on the sidewalk, green-grey from so much loving. we first
have to find better answers than these automatic things. i don't say
this to add responsibilities, but rather so that you know, sister, that
the attempted murder comes from within, like the last refuge of a
cowardly colonialism. come here and i'll give you food and shelter

---

1   From *while they sleep (under the bed is another country)* (Birds, LLC, 2019)
    and first published in *Slice Magazine*.

while i have it, que te añoño, will (cuddle? spoil? hold and rock and
sing?) you, and will duplicate the hugs. i can't heal the fathomless,
but what kind of world would this be without you? what kind of
world is this that harrasses you? without rescue, let's speak of the
future. not as realists, not as visionaries, let's speak of the future
because we will find it in a moth-eaten rug, in the tea of the
drunken, in the *buenos días, there is coffee* of a confused
and sincere embrace. we have a bed and we remember.

yours forever,

raquel[2]

---

2   (nota para una amiga que desea suicidarse después del huracán) nadie nos
enseña a aceptar la muerte porque la muerte, esa muerte de latita, queda vacía
en nosotros: el gran hueco del carajo que nos quiere devorar. nadie nos dice
como podemos integrarnos al nuevo mundo imposible del mañana, como se
supone que evitemos caer en el círculo perfecto de una ojera permanente que
llamamos darle cara al día. mana, ¿cómo no entenderlo? esa es la pregunta
que evito con el fervor organizativo de un equipo de rescate que nunca llega,
pero te voy a decir esto: después del deseo, no siempre viene la muerte. a
veces te encuentro por la calle y brillas como astro o como lámpara solar,
pero igual vales más que todos los generadores (por si no te lo han dicho mil
veces). y otras veces, sin tilde, i.i.i. otras veces, me llegan tus palabras como un
recogido de fondos que explota y temporaliza la verdad, como un aguacate
espachurrao en la acera, verdegris de tanto amar. nos toca primero encontrar
contestaciones mejores que estas mierdas automáticas. no lo digo por añadir
responsabilidades, sino para que sepas que, hermana, el intento de matarnos
viene desde adentro como último refugio de un colonialismo cobarde. vente
pacá, que te doy comida y albergue mientras la tenga, que te añoño y te dupli-
co los abrazos. no podré sanar lo insondable, pero qué mundo sería este sin
tí. qué mundo este que te acosa. sin rescate, hablemos del futuro. ni realistas,
ni visionarios, hablemos del futuro porque lo encontraremos en la alfombra
carcomida, en el té de campanilla, en el buenos días, hay café de un abrazo
confuso y sincero. tenemos cama y memoria.
     tuya para siempre,
     raquel

"I'M QUITE COMFORTABLE"

Abandonment and Resignation after María

Benjamín Torres Gotay

Luis Alberto is about forty years old, a single father, unemployed, who lives in a small wooden house along with his ailing mother, his fourteen-year-old son, and a brother, on the banks of the gorgeous Caonillas River, in Utuado. He speaks in a soft, almost inaudible voice. He answers questions in monosyllables, murmurs, and gestures. He tends to lower his eyes when speaking.

One could guess that he is not the talking type or that he simply was not in the mood to talk the day I met him at his home, at least six weeks after Hurricane María had ravaged his and most of his neighbors' homes. We journalists meet people like that all the time. Not everybody is eager to speak to a notebook-carrying stranger, who, on top of that, is also accompanied by another stranger, this one carrying an intimidating camera. But my guess is that there was something else.

He impressed on me the idea that he was a man utterly overwhelmed by the circumstances of his life. That, certainly, was a part of what I had seen in him. But I also felt that there was something else. There had to be something else.

I never wrote down his last name. I never put him in a story. I spoke to him for no more than twenty minutes. I stopped the attempted interview when it was clear it was going nowhere. We exchanged courtesies and I left, along with my colleague, the photographer/videographer Luis Alcalá del Olmo. We left Luis Alberto's house feeling a bit confused, not understanding at all what we had just heard from him, even feeling a little bit angry with him.

Only half of what had been Luis Alberto's modest house remained; the other half was a pile of debris occupying most of his patio. He had no running water and no electricity (his barrio had to wait more than nine months before getting its power back). His mother had suffered a fall that left her unable to walk. His son, who had ADD and emotional problems, had not had access to his medications since the hurricane.

His desperate situation was very clear to me: the brutal aftermath of María was closing in on Luis Alberto and his family like a giant fist that was about to crush them mercilessly. For food, Luis Alberto and his family depended on charity from an NGO. He gathered water from the nearby river. Neither he nor his brother had a job or any way to earn a living in those incredibly hard post-María days. He depended completely on others' help, and there were not at that critical moment many "others" to offer help deep in the mountains of Utuado, in one of the most isolated parts of the island, where most roads were still blocked by mudslides. He had been abandoned by the state, as were thousands of others in those terrible weeks and months. His life was hanging by a thread.

But when I asked him how he was coping with the dire situation, how he felt, what he expected from the authorities, if he did in fact feel abandoned, how he expected to provide for his son and for his mother, he always gave the same answer: "Estoy bastante cómodo," he said. "I'm quite comfortable."

That's why I stopped the interview.

By the time I met Luis Alberto, I had already spoken with dozens of people affected by María. In the following months, many more interviews would come. I spoke to many people who had lost everything, including their loved ones. Eventually, all of Puerto Rico confirmed what up to that point was already evident: that the state response was grossly insufficient. Most of the people affected by María, as the statistics and official reports show, had not received adequate assistance, if any.[1] In many cases, months had passed before any government help arrived.

They had been, in simple and plain words, abandoned by those who were supposed to take care of them. But despite all that, few of them expressed the feeling that they had been abandoned or that they felt neglected by the authorities. In saying he was "quite comfortable," Luis Alberto may be an extreme example. But his words were not that different from those I heard in many other parts of the island, from people in similar circumstances or even from senior government officials.

For all of them, the prevailing attitude after the storm seemed to be "it could be worse."

That was exactly the case with doña Ida Nieves, a petite seventy-nine-year-old widow, and mother of fifteen, whom I met at her small wooden house in Barrio Marín, in Patillas, an isolated and mountainous community that is one of the poorest on the whole island. Doña Ida lives with one of her sons, who is missing an eye and cannot work. She lives on a monthly welfare check of $180. I went to her house five months after María. There was still no electricity, nor any sign of when it was going to be restored.

Hurricane María ripped off part of her home's roof, which left her with only one bedroom and a living room. Her son had to move his bed to the living room. Doña Ida had to cover her bed in plastic because in Barrio Marín it rains almost every day; the water goes through the tarped roof and soaks the bed. The repairs to her house were estimated at over $7,000. FEMA had given her a little over $2,000.

"I'm OK with that," doña Ida said, showing no anger. "For me, receiving nothing would have been worse. What am I going to do, since I have no job, nor does my son, who has only one eye?" She shrugged and gave a toothless smile.

During this time, I also met doña Eugenia Cruz, eighty-one years old, at her house in Barrio Mameyes in Jayuya. She lives with two of her six children in a patched wooden house that has been lived in for several generations by her late husband's family. The house lost most of its roof, which had been replaced by a tarp that, on sunny days, produced a blue glow that filled the house and on rainy days let the water go through and soak the rooms. FEMA denied them any assistance because, like thousands of families in Puerto Rico, doña Eugenia and her family had no documentation to prove that the land they had inhabited for generations was theirs.

Doña Eugenia's son, Ángel, who was the only person in the household with a salary—if we can call $25 a day picking vegetables at a local farm a salary—said he had no money to repair the roof, nor any chance of getting it. "What I earn is for the basic things," he said. But then again, when asked how he felt about the way he and his family were being treated by the US and Puerto Rican authorities, he responded, "They are doing what they can."

At this point, there is not much left to say about Hurricane María, but no doubt plenty to understand. First, we need to understand the attitude of resignation shown by much of the population in the face of the neglect or outright discrimination by the governments of the United States and Puerto Rico. To begin with, the fact that there *was* neglect has long ceased to be a matter of interpretation. Several official reports from federal agencies, as well as news analysis in Puerto Rico and the US mainland, make it clear that Puerto Rico did not receive the same attention or assistance as Texas, Louisiana, and Florida in the aftermath of major natural disasters.

Perhaps the most important report was published by FEMA itself, in June 2018, in which the agency acknowledged that it was sorely unprepared for María, that it had a shortage of supplies in its Puerto Rico warehouse when María hit, that it failed to deploy enough personnel to the island once the disaster had begun, and that it could not contain the humanitarian crisis that spread throughout the island in the following weeks and months.[2]

I saw this with my own eyes many times: FEMA, military personnel, and, very rarely, Puerto Rico's government officials went to the towns of what we call *la montaña*, in the mountainous center of the island, to deliver aid. But only rarely did they go any farther than the center of town, leaving the vast rural areas unattended, sometimes for weeks. Mayors received the goods, but because multiple roads were blocked by mudslides or because they had no fuel for their own trucks, they could not deliver the supplies to the most isolated areas. The residents of the isolated areas, of course, also had no way of going to town to pick up the goods.

In March 2018 the news site *Politico* published an in-depth article that said that nine days after disaster struck, FEMA had approved $141.8 million in individual assistance to Hurricane Harvey victims in Texas versus $6.2 million to those affected by María in Puerto Rico.[3] The greater Houston area affected by Harvey had at the time of the hurricane a population close to six million people. Puerto Rico had 3.4 million people. Also, during those first nine days, according to the *Politico* analysis, FEMA provided 5.1 million meals, 4.5 million liters of water, and over twenty thousand tarps to Houston residents. During the same period in Puerto Rico, on the other hand, residents received 1.6 million meals, 2.8 million liters of water, and roughly five thousand tarps.

Politico also demonstrated that in the first nine days after Harvey, there were thirty thousand federal personnel in the Houston area, while in the same period only ten thousand had been deployed to Puerto Rico. Also, FEMA approved permanent disaster work for Texas in ten days, compared with forty-three for Puerto Rico.

I myself published a piece in *El Nuevo Día* that proved that Puerto Ricans received assistance that was, on average, inferior to that received by the victims of eight of the eleven most destructive hurricanes since Katrina in 2005, even though María was the third-costliest hurricane to have hit any US state or territory in the last thirteen years.[4]

The assistance received by María's victims in Puerto Rico was less than what was received by the victims of Katrina in Louisiana; Harvey, Ike, Alex, and Rita in Texas; Sandy in New York, New Jersey, Pennsylvania, and Maryland; Irene in North Carolina; and Matthew in Florida. The only US residents who received less monetary assistance than those in Puerto Rico were those affected by Hurricanes Wilma and Irma in Florida.

The difference was substantial, in some cases. For example, those affected by Katrina received on average $9,016, and those affected by Sandy, $9,265. In both cases, the assistance is more than three times the average $2,600 received by the residents of Puerto Rico. According to FEMA, this inequality reflected differences in the cost of living between the US states and the territory of Puerto Rico. Those differences do exist, but they are not substantial enough to explain the inequality in the assistance approved to Puerto Rico's residents, 100 percent of whom were affected, one way or the other, by María.

While this neglect was taking place, Donald Trump, who visited Puerto Rico fourteen days after the storm (he went to Texas twice in the first eight days of their emergency), was fuming publicly about the impact that Puerto Rico's recovery was going to have on the federal budget, feuding with the mayor of San Juan, Carmen Yulín Cruz, and saying things like "Puerto Ricans want everything done for them."[5]

During the first weekend after the storm, when Puerto Rico was coping with some of the most terrible moments of its history, Trump was firing off tweets about his feud with the NFL players who chose to kneel during the US national anthem. According to the aforementioned *Politico* article, those tweets sent "a subtle, yet important message" to the federal bureaucracy: Puerto Rico was not a priority. What happened afterward proved that this analysis was right on target.

It's been said that María confronted Puerto Rico with its ugliest realities, with parts of our own selves we would have never wanted to see. It made visible the supposedly hidden poverty—hidden, that is, for those who don't want to see what doesn't fit their beliefs about who they are. The hurricane unearthed colonialism to its most rotten and despicable core. It showed how alone we are as a society and as a people. It revealed the nakedness of our government institutions and bloated agencies that, despite their enormous budgets, proved unfit to deal with a catastrophe of this magnitude.

And, above all, it revealed the defenselessness and the abandonment of the most vulnerable sectors of our society. It haunts me that María unveiled not only the physical and visible aspects of that vulnerability but also the mental and spiritual solitude that people experienced. This is why it was so hard for many of the people in Puerto Rico to recognize that they had been abandoned and discriminated against. We had been indoctrinated for generations with the notion that there would always be a father figure there to protect us, be it the colonial government of Puerto Rico or the almighty *americano*. Those notions had been hammered into our minds since the dawn of the US invasion, more than a hundred years ago.

In the moment of truth that María thrust upon us, we were left with very little to show for all that we were made to believe for more than a century.

Dumbfounded by such an unexpected turn of events, people could only react with statements of shock like "I'm quite comfortable," "receiving nothing would have been worse," or "they are doing what they can." Not to mention comments from senior government officials, beginning with Governor Rosselló and resident commissioner Jennifer González. Both said that the federal government had given Puerto Rico all it had asked for. Not to mention, either, the scores of people saying publicly that "Haiti would have received zero assistance," which, as we all know, is not only cruel but also very ignorant.

The ugly, unpalatable truth that we were abandoned, that we were forgotten, that we were not important to the US government did not fit, could not fit, in the minds of most people in Puerto Rico. It was inconceivable for them. Hence, María taught us how deep the seed of colonialism is planted inside us and of how little importance we are to the empire.

Perhaps we can say that María, at least, made us understand that we can only count on ourselves.

We'll have to wait until the next crisis to know for sure. I, for one, am not optimistic.

1. Frances Robles, "FEMA Was Sorely Unprepared for Puerto Rico Hurricane, Report Says," *New York Times*, July 12, 2018, https://www.nytimes.com /2018/07/12/us/fema-puerto-rico-Maria.html.
2. FEMA, 2017 Hurricane Season FEMA After-Action Report, July 12, 2018.
3. Danny Vinik, "How Trump Favored Texas over Puerto Rico," *Politico*, March, 27, 2018, https://www.politico.com/story/2018/03/27/donald-trump-fema-hurricane-maria-response-480557.
4. Benjamin Torres Gotay, "Ayuda para la isla tras María fue inferior a otros territorios de EE.UU.," *El Nuevo Día*, March 26, 2018, https://www .elnuevodia.com/noticias/locales/nota/ayudaparalaislatrasMaria fueinferioraotrosterritoriosdeeeuu-2409397/.
5. Amanda Holpuch and David Smith, "Trump Attacks Puerto Rico Mayor: 'They Want Everything Done for Them,'" *Guardian*, September 30, 2017, https://www.theguardian.com/world/2017/sep/30/donald-trump-attacks -puerto-rico-mayor-carmen-yulin-cruz.

NARRATING THE UNNAMEABLE

Eduardo Lalo

In *On the Natural History of Destruction*, W. G. Sebald recalls reading somewhere that the Allied bombing of Dresden created an artificial hurricane. Dresden had no military bases and was not a military
target. Its population was largely civilian. The war was already lost
for the Germans, but Dresden was nevertheless bombed, mainly by
the British. And the number of bombs they dropped on the city was
just incredible. The flames were calculated to be more than a kilometer high; that's almost three-quarters of a mile. Since fire is a phenomenon of oxygen, there were hurricane-force winds within the
city. After the bombing, corpses were found in the public fountains;
it became so hot that people went in them to try to survive. And they
had been cooked, boiled like chickens.

After the bombing, and for many years after that, the city was in
total ruin. Sebald quotes a British journalist who stayed in Dresden
shortly after the war. The journalist said that when he traveled on
the city's tramway, he could tell who had survived the bombing and
who had not. The survivors never looked out the window. I think
that passage is pertinent now and very relevant to what I want to say.
"To narrate something, you have to see it," and one of the things that
happened to us after September 20, 2017, is that for many weeks we
could see very little.

In fact, when I got the first calls from friends outside Puerto
Rico, I asked them, "What is happening?" They were the first ones
to tell us the seriousness of the situation. During a hurricane, you

don't see anything. You see outside your window. That's your horizon at that moment. And if you have to secure yourself in a bathroom as the hurricane passes, you see even less. After that, you only see fallen trees and debris.

The first or second day after the storm, I rode my bicycle from where I live on the outskirts of San Juan to the university, where I have worked for more than thirty years. That's when I saw how it was. Traffic completely chaotic. Lampposts, trees, whatever, in the middle of the street. I arrived at the front of the university, which was closed, and I saw the buildings for the first time, because they had had trees in front before, and now they weren't there. I remember going into Río Piedras, which is the sector of San Juan where the university is, and feeling the silence. It was not because there weren't people around; in fact, there were. But there were no birds, and it's the most uncanny feeling you can have, because that's the sound of death. Everything was dying.

It's difficult to tell the story because from day one whatever came from the government was lies. At first, maybe you do an act of faith in extreme situations, you try to believe whatever they're saying. Very quickly I saw a pattern develop: anyone from the government who spoke publicly, and there were many in the first weeks, always had to mention the governor. I soon realized that even while people had no water, food, or electricity, while people were dying in hospitals and elsewhere, the governor's public relations people and government functionaries had decided to use the hurricane to try to win the next election in three years. Everything its representatives said was later found to be a lie.

So, when I began writing again—my column in *El Nuevo Día*—or answering interviews from journalists in Puerto Rico or elsewhere, I tried to rigorously practice my profession. Literature is a special world. It tries to express tragedy, but the real object of the literary word is the unnameable. You could put into words all the government's irresponsibility and its corruption. And people do that every day. But you cannot use words to capture pain, especially collective pain. Our pain is not only a personal but also a historical pain. It's a pain that is not seen, or that is seen as that of a marginal society whose humanity is undervalued. That's not only because Puerto Rico remains a US colony, but also because our region, the Caribbean, is considered second class. Yet the modern world we know today started in the Caribbean. What we have today is what started October 12, 1492. The Caribbean was the great laboratory for colonialism and imperialism everywhere else. That's why the Spanish conquest of Peru and Mexico was so successful: they knew already how it was done; they had thirty years of experience in the Caribbean to know what worked.

I try not to express what was painful but to use words to point at something that cannot be put into words. The extent of irresponsibility that we have suffered from the government of Puerto Rico, and many of its leaders, and also from the US government. It's impossible to express, it's unnameable. It was recently calculated that during the time when Trump visited for a few hours, in October 2017, at least eighteen people died each day from the hurricane. The governor told him then that a total of only sixteen people had died from the storm, but just in that day, more people died than that. But the tragedy is not only the number of deaths; it's not a contest of numbers. It's the abandonment of a country. The United States used Puerto Rico for its own purposes for more than one hundred years and, in this critical moment, felt that it was in its best interests to leave it behind. That's what colonialism is all about. What is sovereignty but the capacity, .the power, to design yourself? To make your own plans. That's exactly what we don't have.

And American "generosity" is channeled to us according to US priorities, not what is useful to Puerto Rico. We live in a very different situation from that of the fifty states; we are a different country. We are a different nation. To try to put Puerto Rico within the framework of American policy just doesn't work. It's not a question of sending FEMA dollars; it's a question of changing Puerto Rican society to make it more efficient and self-sufficient. And that's not the project of colonialism. The project of colonialism is dependency. This hurricane is being used by government officials in Puerto Rico and the United States to increase that dependency. That's exactly the opposite of being prepared for an emergency. You can send many millions of dollars, but it will be given to American companies, to contractors that are only there for the money. Not to build a country, not to improve society. They will take the money to American banks immediately. We are not receiving more than remedial aid, and we are being used once again.

To tell a story is to discover its real meaning. For me, meaning is encircled here by repetition. We have seen the same thing every day for the past century. Now a whole system of profiteers is being created. Hurricanes have become an economic asset, and that's the sad story of María and Puerto Rico. Of course, there are forces opposing this, but that's the real, less visible story.

IF A TREE FALLS IN AN ISLAND
The Metaphysics of Colonialism[1]

Ana Portnoy Brimmer

> *Ah, what an age it is*
> *When to speak of trees is almost a crime*
> *For it is a kind of silence about injustice!*
> —Bertolt Brecht

but even the trees spoke
snapping like dry bones / under the weighty foot of weather

sparking and slung over power lines
last attempt at sustained flight / sputtering sap and blood

branches a thousand arms reaching / routed / rotting
bark marbled fungus and termite

consider the fright of a flamboyán
skirts upturned

robbed of red / roots ripped
from the earth's scalp

---

1   Approaching a year since hurricane María's passing over Puerto Rico, on
    September 20, 2017.

consider the cry / of a carambola
struck down / stripped bare

starfruit putrid sweet
of the ground / lost aspiring body of the cosmos

consider the plight of a plantain
self-suffocated / wrapped in its own leaf / hijos dying inside

sprouting back like jagged teeth
twisting arthritic fingers / both growing and graves

to not speak of trees now / when even the trees
have spoken / is the deadliest of silences

for when a tree falls in an island / and the world
is around to hear it / the island drowns

for a tree is no longer a tree / and semantics like leaves change
it is memory burnt into body / 7 months piled up on a sidewalk

blocking the street / at your doorstep
blighted wood and abandonment

it is death at its slowest / an unblinking mirror
steadfast rustle / of gutting truth

that if a people fall in an island / and the world is around to hear it
we make a sound—but only the ocean responds / with a swallow

THIS WAS MEANT TO BE
A HURRICANE DIARY

Beatriz Llenín Figueroa

October 20, 2017

1

My idea of keeping a hurricane diary lasted only a short while—about as much time as it took me to remember that I had to fetch water from the well in front of the washing machine repair shop. The municipality's water station was unavailable the three times I tried there.

2

After the hurricane, only the Paleolithic humans who still had Telefónica landlines could call and be called. Telefónica was a public phone company that was privatized by the governor's daddy in the 1990s. Puerto Ricans must adhere to the march of progress led by fathers and sons, right?

The rest of us, privately owned, satellite-controlled creatures, could only report, "I saw her and she's OK." After a brief pause, we added explanatory clauses like:

driving by in the car;

standing in the (eternal) line at the gas station;

exiting the supermarket with two cans of tuna;

at her home, since I had enough gas and her neighbors had cleared the way.

Or, with the anguish of a still unknown loss, we said, and still do, "I have not seen her." To this, we do not add explanations. Can silence explain the agony?

"How are you?" We barely have the courage to say this as we do when we *really* ask that question. Much less do we have the courage to answer. "Good." We lie. We know it. This pain, so oceanic, reduces us to the "good" we can be outwardly, viewed from a safe distance and, especially, considered in proportion to the thousands of people already dead or on their way to dying because of this politico-natural catastrophe.

3

Over this past month, I read a long novel in short spurts stolen from a life besieged by survival. It is a historical novel about the murder of people and socialist utopias, written less than a decade ago. It is also yet another novel narrated from men's perspectives, although women are just as much protagonists as men are in the story of "really existing socialism." Being, quite literally, partners of men is not enough for women to avoid their figuration in the novel as decorative objects, vilified as seductresses or victimized as fools.

The constant presence of dogs is the novel's only redeeming quality, although dogs are not main characters either. The novel is about several men, one of whom loves dogs, and its title closely resembles that phrase. (Apparently, women do not love them. In my house, however, we live according to the exact opposite principle, which is partly why we have remained connected to life. Has someone in government as much as thought about how many dogs have died this past month?)

4

Very little about the hurricane is natural. This is mass murder. Of peoples and utopias, just like in the novel. But Trotsky's assassination no longer needs to be ordered from the other side of the planet. It has also become superfluous to assemble—as happens in other, profuse stories of murdering people and utopias—a complicated structure of spies and informants to carry out a coup d'état in favor of a military strongman who will rule in the US empire's interests. To show that our lives are worthless, it is enough for a hurricane, strengthened disproportionately by the systematic exploitation of the planet led precisely by that empire, to churn atop an ancient colony.

5

A month on, but it has been years. Decades. This is a pain without time. So many of us don't even have the spaces (homes, plazas, community centers) where we used to cry anymore.

This is also a pain that cannot be written. So many bubbles of falsehood have burst that there is no language to encompass them. Even if before the hurricane we knew the lie of Puerto Rico's status as a US "commonwealth," of its self-government, of the gringos' benevolent authority, of the "development" and "progress" that characterize Puerto Rico's socioeconomic and political model, the most colossal bubble burst is the one that allowed me to think—and write—as if "I know all this is a lie." I now understand the enormous vanity entailed in thinking and writing as someone disabused, as someone who knows what is true: our miserable history of enslavement, exploitation, discrimination, dependency, corruption, self-sabotage. After the hurricane, I discover, to my ineffable pain, that the highways, gasoline, cell phones, cement light posts, shiny signs and billboards of tax-exempt corporate chains—that all these things can disguise the misery, and in fact have disguised it for a long time, even though we "knew" they didn't.

6

Before I can finish crying over the presence of soldiers with huge machine-guns in front of a gas station in Mayagüez, I arrive at a store where a very self-confident Boricua citizen howls that *"el boricua está cabrón"* and that in this country we need a few gunshots to get people in line.

I feel so, so much terror.

7

Dear sir in whose opinion *el boricua está cabrón*, if I am allowed to dream that you might read these lines, please, look to the Boricuas working in the Brigada Solidaria del Oeste (West Solidarity Brigade) and producing art on the street affirming that *Borikén florece* (Puerto Rico blooms). When, in one of the Brigada's weekly meetings, I thanked each person I greeted, they all answered, "For what?" Isn't this the "getting in line" we need?

I feel so, so much love.

8

Governor, sir, if you had died, would the death count be, at last, at least, forty-nine?

9

I owe nothing, nothing at all, to the United States of America, nor to its capitalists or any investment firm, nor to the wealthy who get all kinds of tax breaks in this colony cursed by history. It is the United States of America that has built itself on the basis of our misery and, yes, of our death. This death includes our dying for the archipelago because we *have to leave*, against our will.

Thus, I do not want aid because Puerto Ricans are "American citizens." My citizenship is freedom. The gringo citizenship in this country is a war game, a first-world war game, in itself decided on and bestowed at the empire's convenience. The United States of America owes Puerto Rico: *that* is the unpayable debt.

Because of this, because Puerto Rico, like the rest of the Caribbean, has never seen a single cent in reparations for the centuries of exploitation that created the centers of power in Europe and North America, the United States of America has the historical, ethical, and political *obligation* to pay. This is not "aid." It is a debt for which any imperial aid disbursement will always be lacking, for it represents nothing more than an obscene cover-up.

10

To evoke my dear friend Ariadna Godreau Aubert in her book *Las propias*: my supranational aspirations have nothing to do with the United States of Power. On the contrary, my aspirations lie where we ignite our desire to begin from and give birth to so many human and nonhuman lives, bound together by the most abject disinterest for power. Never am I more Caribbean, Antillean, Latin American, wherever we may be, than today. I live and die with wounds that will never finish closing. And I will always stand in the lines we make to care for us and, hopefully, to heal us.

ANOTHER HAPHAZARD GESTURE

Sofía Gallisá Muriente

Author's Note: What follows are partial notes taken in the weeks after the hurricane as part of a useless effort to process the scale of what we lost and what we gained. Some were written on rainy days to soothe the sadness of knowing there were homes that still had no roofs, roads that kept collapsing, and floods that kept recurring. Some were written on Facebook with the help of others who shared experiences and rumors they had heard. Others were written simply to remember conversations.

Some of these lists and images were previously published in 2018 in the zine *Contra Viento y María*, edited by La Piscina Editorial in Tenerife, as part of a project of the Comité de Cariño Criollo, to support Puerto Rican cultural workers recovering from the hurricane. They have been minimally edited.

GONE WITH MARÍA

The Property Registry

The Department of Justice

The Bayamón Court

La Junta (the bar)

The Polygraphic Triennial

Aymat Doughnuts (until the power came back)

The Nuyorican Poet's Café

El Baoricua restaurant

La Chiwinha fair trade store

Fifty-nine employees from GFR Media

Plaza Palma Real in Humacao

The fish farm in Maricao

The boardwalk in La Esperanza, Vieques

La Nasa (the bar, but it reopened next door)

Latte que Latte café (though it reappeared sometime later)

The restaurants Galloway's and El Pirata in Boquerón (but they were both rebuilt)

The plane sculpture in a rotunda in Guaynabo

The space rocket in the Science Park in Bayamón

Part of the collection of Puerto Rican art from the library of the School of Arts and Design

The ficuses in the square of Old San Juan where people played dominoes and parrots nested

The parrots that used to screech in front of the University of Puerto Rico's Río Piedras campus

The archives of Remi the clown

The professional basketball and volleyball seasons

Las Justas (the college sports championship)

Fresh fruit (for a while)

The beaches of Condado, Ocean Park, and Escambrón (which were toxic for months)

The restaurant El Nilo in Río Piedras (it came back!)

The Natatorium

The Institute of Puerto Rican Culture's Office of Patrimony

The Santiago Iglesias Pantín Labor Archive at the University of
    Humacao
Sistema TV
The Vico C concert
The Ricardo Arjona concert (at least some good news)
The sales tax on food (for a limited time)
La Garita bar in La Perla (but they rebuilt bigger and stronger)
Whatever dignity the Supermax grocery in Old San Juan had
    left
The expert witness in my mom's robbery case
The print edition of *Claridad* newsweekly (for a couple of issues)
The chimney of an abandoned sugar plantation next to the
    Caguas Walmart
The solar panels from Paseo Puerta de Tierra and Cuartel de Ballajá
The motel El Panamericano in Hatillo
An antenna on the radio telescope in Arecibo
The caves in Camuy (for at least thirty years)
Punta Ballena beach
Two million chickens

THINGS I'M HAPPY MARÍA TOOK
Fences
Billboards
Stoplights
My schedule
"Normality"
The shame in being sad

LOST TO MARÍA

Marilú's mom
Martín's mom
Alicia's mom
Rubén's mom
Belisa and Luis Alberto's dad
Luis's dad
Pablo's uncle

BARAHONA, MOROVIS

The night Alice's dad died, he had to describe his stroke symptoms over the phone to 911 so they would pick him up, after they ignored his daughter's calls. He had been released that morning from a hospital running on generators that was rationing oxygen, during the post-Irma, pre-María blackout. Alice accompanied him to the hospital, and at the very moment the old man died, his grandson Yandel, who was at home, became pale and vomited, and a flock of bluebirds flew out from among the *mogotes* (hills). Alice waited in the hospital for her brothers, but they trusted that she would call them if something was wrong. The phones weren't working, so no one found out in time to say goodbye. The nurse waited with Alice until it was late. In the funeral home, Alice's sons were scandalized when they saw their grandfather with makeup on. "He would never do that. He's not a fag."

WHO'S WORSE?

People in San Juan say people elsewhere on the island are the ones who are really suffering. People elsewhere on the island say that things are worse in San Juan than in the countryside, because of the stress. Everyone thinks Vieques has collapsed, but a friend who lives there says there are assemblies in the public square every day and that a young woman got up at one of them to yell at the mayor for walking around drunk while alcohol sales are prohibited. I've also heard power generators were stolen at the morgue and the hospital. How do you hide one of those?

A month after María, in Sabana Grande, there was power, water, internet, well-stocked supermarkets, clean streets, open Chinese ice-cream shops, and even Halloween decorations.

In San Sebastián they started their own power authority with retired electricians and figured a way out.

In Adjuntas they're still installing solar panels, for the next one.

In El Yunque National Forest there's a road where a couple of utility poles fell down that supplied electricity to a neighborhood, but now the federal government doesn't want to restore them because they're in a natural reserve.

In Utuado there are neighborhoods that were land occupations and still have no power. Some say they're trying to get people to leave to set up Airbnbs.

In Old San Juan and Puerta de Tierra the population of addicts and homeless people has grown.

I want to spend the next hurricane in Cabo Rojo. The mayor there went door to door loaning his satellite phone to people so they could call their families and tell them they were all right.

QUESTIONS

I've learned that when saying hello, you don't have to ask *How are you?*

I've also learned that, sometimes, to answer the question *Who won the game last night?* you have to check your phone casually, as if keeping it charged and getting a signal weren't a big deal.

MARÍA DAYS

100 percent probability of rain (it's one of the few certainties),
Strong and constant twists and turns,
Gusts of wind, bursts of sounds, in the body. *Ráfagas.*
The calm eye of the storm passed by a long time ago, but the hurricane hasn't left.
Where it floods once, it floods ten times.
Where power comes back, it might not stay.
Where a bridge has collapsed, sometimes chairs and a bonfire pop up.
Where people haven't seen FEMA, they've seen friends, relatives, curious strangers, the press, the diaspora, and the municipal government.
Where there once was, now remains.
From the sky, we are back to looking green.
At night, we're barely visible.
I heard we're in a struggle against the light,
the rupture won't last, we must plant seeds.
I also heard that we should burn all flags.
"There's something gone from the mountains," says Lorraine.
"We have to study New Orleans," say people at La Cucina.
"There's multiplicity, but we're here," says Nibia.
"There are things you have to let collapse," says Gilberto.
"You need a rock-solid stomach to deal with this country,"
said my dad, since before the storm.

PART III

Representing the Disaster

OUR FELLOW AMERICANS

Why Calling Puerto Ricans "Americans" Will Not Save Them[1]

Frances Negrón-Muntaner

In September 2017, the category 4 hurricane María obliterated Puerto Rico's electrical grid, destroyed or damaged half a million homes, and killed at least three thousand people. Nearly four months afterwards, close to half the island's households remained in the dark, and one hundred and fifty thousand people were forced to leave. A year later, partial blackouts remained common, and 4 percent of Puerto Rican households still did not have regular access to electricity. This prolonged darkness, however, has clarified what over a century of legal euphemisms have tried to obscure: that although US citizenship was imposed on Puerto Ricans in 1917, living as nonvoting citizens on a colonial possession that "belongs to, but is not part of" the United States not only spells disaster.[1] It can also spell death.

While many Puerto Ricans have met this realization with a mixture of outrage, resignation, and indifference to the United States, some Americans have responded with a historically rare gesture: rhetorical incorporation. As the hurricane approached, mainstream media outlets started to describe Puerto Ricans as "3.4 million

1   Originally published as Frances Negrón-Muntaner, "Our Fellow Americans: Why Calling Puerto Ricans 'Americans' Will Not Save Them," *Dissent*, January 10, 2018.

American citizens living in Puerto Rico." Once the storm devastated the island, the federal government failed to fully assist, and President Trump disparaged Puerto Ricans, journalists and other public figures started to entirely drop the distinction between citizen status and national identity. Perhaps echoing Governor Ricardo Rosselló's mournful address to "my fellow Americans" on the eve of María's landfall, they started to simply call Puerto Ricans "Americans" and, even more unexpectedly, "our fellow Americans."

One of the first outlets to do so was the Weather Channel, whose anchors and reporters, especially Paul Goodloe, repeatedly declared that Puerto Ricans are Americans. CBS correspondent David Begnaud, who emerged as one of the most compelling and dedicated journalists of the immediate post-María crisis, routinely did the same. In an October 5 report on island conditions, for example, he stressed with exasperation, "*These are Americans* sitting in line, sleeping in their cars, desperately trying to get fuel."[2]

MSNBC commentator Rachel Maddow was at times more emphatic. Reporting on the infamous "spa day" story, in which a US doctor quit after witnessing medical personnel turn a hospital triage center into a manicure station, Maddow remarked in an alarmed tone: "You have people starting to die, *Americans starting to die*, in Puerto Rico because of treatable bacterial infections . . . This storm is no longer killing Americans—the federal government's response to the storm is now killing Americans."[3]

For a number of US journalists, continually calling Puerto Ricans "American" was a way to keep audiences following the story and to maintain the political pressure on the federal government. In an unusual case of reflexivity and advocacy in a mainstream publication, reporters Kyle Dropp and Brendan Nyhan of the *New York Times* even made the case that what "three million Americans in Puerto Rico" were termed was a life-and-death question and underscored the link between rhetoric and survival.[4] Commenting on a poll about US perceptions of Puerto Ricans shortly after the hurricane struck, they noted that Americans in general—including Trump supporters—were more comfortable with the idea of providing aid to Puerto Ricans once they knew that they were US citizens. The finding was reported as both a conclusion and a plea: "Our sympathies for other people depend in part on whether we see them as fellow members of our tribe. Without more coverage, it may be easy to forget that the people suffering are our fellow Americans."[5]

Journalists were not alone in embracing rhetorical incorporation as a strategy. Corporate America also got in on the act. Perhaps the most striking instance was a public service announcement produced by the megastore Walmart, which began running on primetime television on October 9. Called "United," the PSA was not Walmart's first on a hurricane-related theme. The company had released another PSA a month earlier to raise funds for those affected by Hurricane Harvey in Texas. The thirty-second Texas spot began with a photograph of Houston partly underwater and ended with a written message: "To the Lone Star State, you are not alone." According to adland.com, this message helped raise over $25 million.[6]

The Puerto Rico English-language version was similar in form and included references from American popular culture, suggesting that in the eyes of Walmart the island was equal to any state. The video opened with images of the hurricane's impact under an instrumental arrangement of Ben E. King's classic "Stand by Me." Although the lyrics are not part of the PSA, the song was likely chosen to allude to the destruction of the power grid and ensuing darkness: "When the night has come/And the land is dark/And the moon is the only light we'll see/No, I won't be afraid/Oh, I won't be afraid/Just as long as you stand, stand by me." In this rendition, a male voice-over heard during the spot's duration declares, "To our fellow Americans in Puerto Rico, we may be separated by an ocean, but we are united."[7]

US politicians eventually joined in too. In an October 16, 2017, address to the legislature after a visit to Puerto Rico, Florida Democratic senator Bill Nelson repeated the phrases "our fellow Americans" or "fellow citizens" six times, ending with "Our fellow Americans are dying, and they desperately need our help. . . . I'm here to urge this Congress and the administration that we have to act."[8] While Puerto Ricans in general, including elected public officials, used the phrase significantly less often, New York Congress member José Serrano also pressed his House colleagues the following week in analogous terms:

"[The people of Puerto Rico] are our fellow Americans; they've served in our wars, they do pay taxes, and they should be treated equally."[9]

The rhetorical explosion of "our fellow Americans" raises the question of why this phrase emerged and why it arose with such force. Part of the answer lies in how the term "American" remains a signifying site for struggles over what and who Americans are (or can be), and who may claim full national membership and access the benefits of US citizenship. Although Puerto Ricans are mostly ignored in these debates, US journalists and political actors have occasionally called attention to Puerto Rico's colonial subjection through the vocabulary of citizenship and national belonging. A portion of a 1900 *Washington Times* article quoted in the *Congressional Record* offers a succinct example that feels thoroughly current in both content and rhetoric. After the House of Representatives voted on legislation asserting that the nation's constitution did not apply in full to the island, the *Times* proclaimed, "The crime against our suffering fellow-Americans in Puerto Rico, and against the Constitution of the United States, is complete."[10]

Another part of the answer lies in the idiom's relationship to presidential rhetoric. The phrase stems from "my fellow-citizens," in use since the republic's early period and employed by the nation's first president, George Washington. The present usage consolidated in the twentieth century through its reiteration by various presidents, including Franklin D. Roosevelt, who in 1933 was the first to adopt it in an inaugural speech. The expression's contemporary resonance, however, can best be traced to Lyndon B. Johnson, who integrated it into every State of the Union and other key speeches on divisive subjects in which American lives were at stake. Among these were Johnson's last two addresses to the nation. On each occasion he began with "Good evening, my fellow Americans," before announcing that he would first limit, then halt bombing in Vietnam, and not seek reelection.

At a time when the president knowingly failed to safeguard US citizens, the phrase's association with presidential addresses and their effect of rhetorically constituting the "American people" propelled its surge following María. The repetition of "our fellow Americans" affirmed that Americans themselves have the power to incorporate Puerto Rico as part of the national body politic and dictate that Puerto Ricans deserve care as US citizens. In other terms, to the extent that President Trump has explicitly claimed the category of "American" for whites, "our fellow Americans" seeks to refuse the president's racist rhetoric by taking over presidential speech. The phrase ultimately aims to be a declaration of democratic inclusion—the chosen version is, after all, "our" and not "my" fellow Americans—and a form of protest against the explicitly racist rhetoric of the nation's head of government.

In retrospect, this strategy has had tangible effects. Brigades of nurses, pilots, cooks, celebrities, and even the *Simpsons'* cartoon bartender Moe have sent money and supplies to Puerto Rico or donated their time. A Kaiser Family Foundation poll found that a month after María hit, 62 percent of Americans agreed that Puerto Ricans were not getting the help they needed.[11] More notably still, according to a Fox News poll, the percentage of Americans who supported admitting Puerto Rico as a state of the union jumped, from 30 percent in 2007 to 41 percent in 2017.[12] If the numbers are reliable, this is a remarkable outcome, since as a result of the expanded coverage, it became widely known that the island is experiencing a massive debt crisis hinging on $120 billion in bond and pension obligations, that it will require billions of dollars to recover from the debt and hurricane devastation, and that nearly half the population lives below the poverty line.

At the same time, the deployment and embrace of "our fellow Americans" is significantly more vexed than it appears. A good example is Walmart's PSA, mentioned above. While the company may have intended to raise money for relief, many Puerto Ricans saw the ad as hypocritical given the company's impact on the island. Since 2000, Walmart has become Puerto Rico's largest private employer and its highest-grossing retailer, with upwards of $2.75 billion in annual revenue and more stores per square mile than anywhere else on earth.[13] In the process of attaining market dominance, Walmart has destroyed the vast majority of local competition, including independent pharmacies, and has been a factor in over eight hundred small business bankruptcies. Its near monopoly has likewise driven up unemployment: for every new job that Walmart creates, the economy loses an average 2.3

positions. Equally important, the types of jobs that Walmart has generated are mostly part-time, minimum-wage ones, without benefits or the right to organize. In this regard, Walmart's apparent post-María largesse masks its predatory relationship to Puerto Rico.

But even for institutions, companies, or groups without Walmart's history, the rhetorical gesture of incorporation remains complicated. Although recognizing Puerto Ricans as Americans may translate into greater attention in some contexts, the strategy of "liberal generosity," as philosopher Nelson Maldonado-Torres termed it, is problematic for what it assumes, conceals, and omits.[14] Not only has the rhetoric failed to disrupt colonial relations, it also reveals and entrenches colonialism in fundamental ways.

For instance, those who insist that Puerto Ricans are "our fellow Americans" seek to address Americans' lack of knowledge about Puerto Rico; according to one poll, only 54 percent of Americans were aware that Puerto Ricans are legally US citizens in late 2017[15]—one hundred years after Congress passed the Jones-Shafroth Act that conferred US citizenship to island residents. Yet, in attempting to convey that Puerto Ricans are US citizens, commentary and coverage tended to obscure the coloniality of such citizenship. In fact, the appeal's relative success in generating empathy is possible solely because the vast majority of Americans do not know (or care) that Puerto Ricans are not equal citizens of the United States but territorial citizens with limited rights. In this regard, "our fellow Americans" can only be a viable media strategy if it veils that Puerto Rico is a colonial possession and its citizens colonial subjects. As writer Eduardo Lalo put it, for Puerto Ricans, "citizenship is a trap" that allows the United States to erase or avoid acknowledging that its occupation of Puerto Rico is an act of colonialism.[16]

Rhetorical incorporation also ignores the intimate ties between colonialism and racism, and it sidesteps that citizenship has never offered full protection for racialized, colonial, and otherwise minoritized legal citizens in the United States. In this way, the rhetorical gesture reaffirms the myth that once citizenship is acquired, all Americans experience it equally, which is not historically nor currently the case. This is dramatically evident in recent federal hurricane response: US citizenship did not protect African Americans from dying during Katrina (2005) and has not kept Black and Latinx communities stateside from being flooded with toxic waters, which will harm "Americans" for generations, in the aftermath of Hurricane Harvey (2017). Neither the state nor citizenship status alone has ever guaranteed rights for racialized people.

The deployment of "our fellow Americans" likewise exposes how US nationalist identification and subjection, rather than the acceptance of colonial responsibility or a shared humanity, is presented as the only way to generate support for Puerto Ricans. Maldonado-Torres has explicitly critiqued the "American" appeal by pointing out that Puerto Ricans should "receive help because they are people, not because they are [American] citizens."[17] In this context, the trope reiterates US rhetorical sovereignty by repositioning Americans, generally white Americans, as sources of authority and "authenticators" of Puerto Rican humanity solely via *their* acknowledgment of the group's Americanness.[18] US political actors presume that this recognition is an intrinsic good and fail to consider that the majority of Puerto Ricans may desire to be recognized otherwise.[19] "Our fellow Americans," then, also erases Puerto Rican ethnic, national, and other identities, as well as their long history of struggles against US colonialism.

Yet perhaps the most critical aspect of the attempt to rhetorically make Americans out of Puerto Ricans is its utter failure to end the suffering on the ground. This outcome highlights that Puerto Rico's colonial subjection is thoroughly systemic, involving multiple politicial, legal, economic, and discursive relations of power favoring US capital and elites. Despite the rhetoric of inclusion, four weeks after the hurricane, no fewer than sixty-nine House Republicans voted against additional aid for Puerto Rico.[20] Usual responses to disasters in the United States, such as expanding the food stamp program for families in need, were not followed because, unlike in the fifty states, there is a cap on the amount of funds the island can receive, even in times of emergency, due to its territorial designation. For still-disputed reasons, the mutual aid typically provided by states when catastrophe hits also did not kick in. States like New Jersey, Virginia, and Massachusetts were quick to send personnel, equipment, and other assistance to Texas and Florida following Hurricanes Harvey, Irma, and María, but not to Puerto Rico. Consistently, the quantity and quality of journalistic interest itself largely continues to replicate existing colonial relations. After María there was a spike in reporting about Puerto Rico, which was more extensive than previous coverage of the island's debt crisis. This expanded reporting, however, has been considerably less than that accorded to hurricane damage in Texas and Florida as María "received only a third as many mentions in text as hurricanes Harvey and Irma."[21] The equally devastated US Virgin Islands—where three-quarters of the population is Black and which has a relatively small and unconcentrated diaspora—received nearly no coverage at all. As noted by researchers Anushka Shah and Allan Ko, and journalist Fernando Peinado, attention to Puerto Rico picked up only after Trump visited the island and has faded since. In addition, whereas media coverage in the United States focuses on the needs of residents and the hurricane's impact on families, Puerto Rico's coverage is largely related to partisan politics, debt, and taxes,[22] underscoring that media interest is often motivated by anti-Trump or anti-Republican sentiment rather than a commitment to resolving urgent problems, ending US colonialism, or addressing climate change in the region. Millions of people seem more outraged by Trump's tossing paper towels at Puerto Ricans than US colonial rule in Puerto Rico, the Virgin Islands, and other US territories.

In other words, the liberal appeal of "our fellow Americans" is no match for a racist, colonial logic designed to keep resources away from racialized "minorities" deemed undeserving. Trump personally underlined this logic when he defended his administration's negligent response by invoking enduring stereotypes of Puerto Ricans as lazy, responsible for their own financial woes, and unreasonable in their expectations of the US government. In contrast to his September 1 speech on post-Harvey Texas, when he stated, "We will support you today, tomorrow and the day after. . . . We help our fellow Americans every single time," Trump's October 12 tweet to Puerto Ricans claimed that "we can't keep FEMA, the Military & First Responders . . . in P.R. Forever." Predictably, this assertion is inconsistent with FEMA's record; the agency sometimes stays engaged in disaster areas for more than a decade after a storm.[23] In fiscal year 2017 alone, FEMA was slated to provide $440 million in relief to Gulf Coast states for damages resulting from 2005 hurricanes Katrina, Rita, and Wilma, and $1.4 billion to New York and New Jersey for Sandy. A year since María, FEMA maintains a presence in Puerto Rico, although its impact appears to be minimal.

But even if the federal government were handling Puerto Rico's crisis like one in any other US state, the island would not be spared another, similarly harsh "American" response: disaster capitalism. Well-connected companies receive no-bid multimillion-dollar contracts to rebuild at an exorbitant cost;[24] recovery gives way to land grabs, privatization, and more debt. Repeating what happened after Katrina to majority-Black New Orleans, whose citizens had full congressional representation, the interests of most Puerto Ricans are being sold out with impunity. Moreover, as befits colonial governance, a provision of the Republican tax bill passed by Congress in December 2017 took this profiteering to new heights. Though the final bill didn't include a threatened 20 percent excise tax on goods imported to the mainland—which could have cut as much as one-third of the local government's revenues—a separate 12.5 percent tax on income from intellectual property could gut Puerto Rican pharmaceutical manufacturing, one of the largest and best-paid sectors of the island's economy.[25]

Ultimately, in this unnatural disaster made by the intersection of climate change, modern colonialism, and Trumpism, it is evident why an appeal to "our fellow Americans" has not saved Puerto Ricans. For too many, including those who now control the nation's resources, the phrase continues to fundamentally refer to privileged whites. Neither legal nor rhetorical incorporation undoes the fact that citizenship is a category that goes beyond legal status, that relegates minoritized groups to an infra-citizenship, and that confers only tenuous claims to American identity and the protections it presumably offers. Paradoxically, the longer the US government deliberately fails in Puerto Rico, the more fleeing Puerto Ricans will legally turn into full-fledged US citizens, further unsettling Americanness.

1. For a more extended exploration of the trope of darkness after María, see Frances Negrón-Muntaner, "Blackout: What Darkness Illuminated in Puerto Rico," *Politics/Letters*, March 2, 2018, http://politicsslashletters.org /blackout-darkness-illuminated-puerto-rico/.

2. David Begnaud, "Puerto Rico Hurricane Damage," *CBS News*, October 5, 2017, https://www.cbsnews.com/video/puerto-rico-hurricane-damage -david-begnaud/.

3. Rachel Maddow, "Doctor Quits Puerto Rico Medical Relief Team over 'Spa Day,'" *Rachel Maddow Show*, October 12, 2017, http://www.msnbc.com /rachel-maddow/watch/doctor-quits-puerto-rico-medical-relief-team-over -spa-day-1072301123804.

4. Kyle Dropp and Brendan Nyhan, "Nearly Half of Americans Don't Know Puerto Ricans Are Fellow Citizens," *New York Times*, September 26, 2017, https://www.nytimes.com/2017/09/26/upshot/nearly-half-of-americans -dont-know-people-in-puerto-ricoans-are-fellow-citizens.html.

5. Dropp and Nyhan, "Nearly Half of Americans."

6. "Walmart 'That's Texas' (2017) :30 (USA)," posted by "kidsleepy," September 2, 2017, Adland.com, https://adland.tv/commercials/walmart-thats -texas-2017-30-usa.

7. "Walmart Puerto Rico Relief Fund: 'United,'" October 9-31–2017, https:// www.youtube.com/watch?v=c9pBgB3dllo.

8. Bill Nelson, "'Our Fellow Americans are Dying, and They Desperately Need Our Help,'" *Sunshine State News*, October 17, 2017, http://sunshinestatenews .com/story/bill-nelson-takes-senate-floor-speak-about-puerto-rico-our-fellow -americans-are-dying-and-they.

9. Bonnie Castillo, "'As Puerto Rico Suffers, Where Is Our Government?' Nurses Join Congress Members to Urge Action on Healthcare Crisis," National Nurses United blog, October 27, 2017, https://www .nationalnursesunited.org/blog/puerto-rico-suffers-where-our-government -nurses-join-congress-members-urge-action-healthcare.

10. 56th Cong. Rec. H3378 (March 27, 1900).

11. Sandra Lilley, Suzanne Gamboa, Daniella Silva, and Carmen Sesin, "'When Did We Stop Being America?' Puerto Ricans Angry, Dismayed over Trump Tweets," *NBC News*, October 12, 2017, https://www.nbcnews.com/storyline /puerto-rico-crisis/when-did-we-stop-being-america-puerto-ricans-dismayed -over-n810151.

12. "Fox News Poll: Support for Puerto Rican Statehood Increases in Wake of María," *Fox News*, October 26, 2017, http://www.foxnews.com/politics/fox -news-poll-support-for-puerto-rican-statehood-increases-in-wake-maria.html.

13. Joel Cintrón Arbasetti, "Puerto Rico First in the World with Walgreens and Walmart per Square Mile," Centro de Periodismo Investigativo, May 7, 2014, http://periodismoinvestigativo.com/2014/05/puerto-rico-first-in-the -world-with-walgreens-and-walmart-per-square-mile/.

14. Nelson Maldonado-Torres, respondent remarks, "Aftershocks of Disaster: Puerto Rico a Year after María," Rutgers University, n.d., https://livestream

.com/rutgersitv/aftershocks/videos/180935036.

15. Dropp and Nyhan, "Nearly Half of Americans."

16. Eduardo Lalo, "Aftershocks of Disaster: Puerto Rico a Year after María," Rutgers University, n.d., https://livestream.com/rutgersitv/aftershocks /videos/180935036.

17. Facebook post, September 27, 2017, and personal communication, December 28, 2017.

18. For a discussion of the concept of "rhetorical sovereignty," see Scott Richard Lyons, "Rhetorical Sovereignty: What Do American Indians Want from Writing?," *College Composition and Communication* 51, no. 3 (February 2000): 447–68.

19. This is consistent with US media portrayals of Latinos more generally, portrayals that, as Félix Gutiérrez has written, seek "to cover or portray Latinos to a largely Anglo audience through mass entertainment and news media with images, issues, and stories that will appeal to and attract that audience." They also tend to represent Latinos "as weaker or less engaged people needing the help of Anglos to make progress." Félix F. Gutiérrez, "More Than 200 Years of Latino Media in the United States," American Latino Theme Study, National Park Service, US Department of the Interior, n.d., https://www .nps.gov/heritageinitiatives/latino/latinothemestudy/media.htm.

20. Cristina Marcos, "69 Republicans Vote against Aid for Puerto Rico, Other Disaster Sites," *The Hill*, October 12, 2017, http://thehill.com/blogs /floor-action/house/355225-69-republicans-vote-against-puerto-rico-aid.

21. Anushka Shah, Allan Ko, and Fernando Peinado, "The Mainstream Media Didn't Care about Puerto Rico until It Became a Trump Story," *Washington Post*, November 27, 2017, https://www.washingtonpost.com/news/posteverything/wp/2017/11/27/the-mainstream-media-didnt-care-about-puerto-rico-until-it-became-a-trump-story/?utm_term=.8d5cd79ada7b.

22. Shah, Ko, and Peinado "The Mainstream Media Didn't Care."

23. Ryan Struyk, "FEMA Actually Can Stay in Puerto Rico Indefinitely," CNN, October 12, 2017, http://www.cnn.com/2017/10/12/politics/fema-trump-hurricane-puerto-rico/index.html.

24. Vann R. Newkirk II, "The Puerto Rico Power Scandal Expands," *Atlantic*, November 3, 2017, https://www.theatlantic.com/politics/archive/2017/11 /puerto-rico-whitefish-cobra-fema-contracts/544892/.

25. Armando Valdés Prieto, "How the GOP Tax Bill Will Wreck What's Left of Puerto Rico's Economy," *Washington Post*, December 20, 2017, https://www .washingtonpost.com/news/posteverything/wp/2017/12/20/how-the-gop -tax-bill-will-wreck-whats-left-of-puerto-ricos-economy/?utm_term= .360324ff7cab.

US MEDIA DEPICTIONS OF CLIMATE MIGRANTS

The Recent Case of the Puerto Rican "Exodus"

Hilda Lloréns

The Caribbean is not an idyll, not to its natives. They draw their working strength from it organically, like trees, like the sea almond or the spice laurel of the heights. Its peasantry and its fishermen are not there to be loved or even photographed; they are trees who sweat, and whose bark is filmed with salt, but every day on some island, rootless trees in suits are signing favourable tax breaks with entrepreneurs, poisoning the sea almond and the spice laurel of the mountains to their roots. A morning could come in which governments might ask what happened not merely to the forests and the bays but to a whole people.

—Derek Walcott[1]

In the first weeks after Hurricane María made landfall in Puerto Rico, the media circulated images depicting a sea of Brown and Black bodies crowding the Luis Muñoz Marín International Airport, the massive number of expectant passengers overflowing out to the airport's curb. The island's largest point of entry and departure was at a standstill because, like the electric grid, telecommunications equipment, radar, and other navigational aids had been destroyed by the category

4 storm. This privatized airport underwent multimillion-dollar renovations in 2014 and did not sustain major damage.[2] And yet commercial aviation's complete dependence on electricity, computers, and telecommunications technology made it nearly impossible for flights to land or take off in the first days after the hurricane.

In the photographs and videos, the people at the airport look tired, their faces registering a range of emotions from consternation and dejection to outright despair. Some sat upright, but many were slumped in chairs. Others were lying down, resting or fast asleep on the airport's shiny concrete floor, their luggage acting as makeshift chairs, beds, and pillows. Still others stood in snaking lines, exhausted babies and toddlers carried by family members, as elderly and frail-looking people in wheelchairs lined up in the departure gates. "After Hurricane María, Airport Looks Like a Shelter," read a headline. "Puerto Rico's Biggest Airport Becomes a Refugee Camp," read another.[3] A JetBlue clerk told a reporter, "People prefer to wait here instead of dying on the island."[4]

This airport-cum-shelter offered little physical respite or comfort to the stranded passengers. Instead, it was miserably hot; temperatures inside its walls reached 100 degrees Fahrenheit.[5] The airport's design, like most "modern" buildings on the island, is constructed with steel, concrete, and glass, making it entirely dependent on air-conditioning to cool its cavernous rooms. But when the electricity that powers the massive cooling system fails, ventilation halts. The airport's fortress like design also makes it reliant on artificial lighting. The twenty-one generators used to power the airport could barely produce enough electricity for a few scattered lights and industrial fans. To boot, these generators depended on a rapidly dwindling and dangerously low fuel supply.[6] Such concrete buildings can, for the most part, withstand the harsh cyclonic conditions that periodically sweep through the Caribbean's "continent of islands," but when the energy-hungry systems that sustain them fail, they quickly turn into dark, dangerously hot heat traps.[7]

It was reported that the people camped out at the airport were desperate to escape the chaotic conditions wrought by the hurricane. One woman told a reporter, "I just don't feel safe here," and by "here" she meant the place beyond the airport.[8] Another commented, "We are running for our life" and "Once we got into the gate, I was finally able to calm down because I knew we were leaving."[9] The airport, it seems, became a de facto place of refuge from the catastrophe still unfolding throughout the island; perhaps people felt strength in numbers, or an imagined closeness to *alla fuera* (out there), as Puerto Ricans often refer to the continental United States, where life was imagined as "safe," "orderly," and "normal."

The scenes at the airport, of "throngs" of Puerto Ricans at the ready to board planes bound for the continental United States, became a focal point in the news cycle, as well as in the greater construction of a biblically tinged post-María narrative dubbed as "the Puerto Rican exodus." Following the biblical etymology of *exodus*, Puerto Ricans bound for the continental United States were seeking "salvation" after experiencing the theophany brought upon them in the form of a merciless hurricane that violently stripped them of normality and, for some, the possibility of continuing to make a life on the island. Underscoring the fleeting nature of time, and possibly of life itself, a twenty-six-year-old man who said he was thinking about migrating told a reporter in October 2017, "Nothing is telling us everything will be OK in one or two years. We don't have that time to waste."[10] In the same article a medical doctor explained, "The easy way out will be to buy a ticket and head out." A fifty-eight-year-old woman who was standing in a long line to get ice made an ominous prediction: "Everyone will leave—everyone."[11]

The airport opened on a limited schedule just two days after the hurricane, slowly ramping up commercial flights in the weeks that followed. The media reported that thousands of Puerto Ricans filled each airplane to capacity with seemingly no end to the numbers of people departing daily. This exodus had no messianic or charismatic leader at its helm. Rather, the group's organizing logic rested on the collective realization, represented in the number of people leaving the island, that for them the present was untenable. For many, journeying away from a devastated home or from the wrecked island or from an unimaginable future in the homeland meant stepping into an uncertain future in the continental United States.

VISUALIZING THE "DISASTROUS TROPICS"

The images from the airport peddled several stereotypes about the *disastrous tropics*. I use this term to index a racialized historical trope in the Western canon, a history produced in the world's "temperate" zone, casting tropical nature, including its Indigenous (Brown) and African (Black) inhabitants, as fecund, unrestrained, and unrelenting, capable of great ferocity.[12] The crowds at the airport stood as evidence of the fecundity of tropical people, who often appear gathered in intergenerational groups. A key characteristic of this trope is the number of babies and children present, easily cohabiting with each other and with tropical nature, including vegetation and nonhuman animals.

Imagery from the disastrous tropics quintessentially depicts individuals, groups, and crowds who look variously stunned, bewildered, dejected, destitute, traumatized, unfit, ill, disabled, dying, or dead. To complete this tableau, individuals or crowds must be surrounded or engulfed in any or all of the following: dirt, dislodged earth, mud, debris, squalor, ruins, stifling heat, mosquitoes, flies, dirty water, and the implicit stench of decomposing bodies, either human or animal. To complement the picture of societal upheaval and disarray, the visual and media narrative emphasizes the absence of order, of policing and the state, the lack of goods and resources, and pleas for humanitarian aid and relief—which is almost always followed by news of a lengthy period of dystopian uncertainty and spotty reconstruction efforts, mismanagement of aid, funds, and botched projects, all confirming the trope of the ungovernability of tropical nature and its inhabitants.[13]

This trope implies that because the tropics are so impetuous, their people live by default at the edge of disaster. To be livable for "modern," "evolved" people, tropical nature must be tamed using human ingenuity and advanced technology. For instance, electrification extends the day by lighting the dark tropical nights; the air-conditioner makes bearable the stifling, worsening heat of a changing climate; and air travel closes the gaps in distances, allowing people to move quickly across the globe. In the Western imagination technological advancements and innovations make the tropics habitable; modern technologies like electrification, vaccination, and water purification, to name a few examples, are all central in the racial matrix of Western imperial power. Touted as helping citizens of the Global South join global modernity and obtain progress, "neutrally" benevolent technological innovation is so naturalized as to hide embedded beliefs about the North's superior scientific intelligence and the South's inferiority.

When disaster strikes, the lack of essential technologies becomes a matter of life and death. In Puerto Rico, an American colony that depends much more than its Caribbean neighbors on electricity generated from fossil fuels, the lack of access to alternative energy sources, such as modular solar panels to power hospital operating rooms or dialysis machines, for instance, led to preventable deaths.[14] Yet many of these very "advances," such as electrification and airplane travel, depend almost entirely on fossil fuels, and they are among the top sources of greenhouse gasses, which have led to an increasingly unbalanced nature that is now producing stronger, more violent Atlantic hurricanes, such as María. Unjustly, island societies and low-lying coastal areas in the Caribbean and Pacific are on the front lines of climate change and will likely continue to experience its worsening effects and consequences, even though these societies contribute only a tiny fraction of all greenhouse gas emissions.[15]

After Hurricane María, the media produced thousands of images of Puerto Rico and its people confronting a destroyed island, leaving it, or living in crowded hotel rooms in various US cities, and these images now join the last century's visual archive of catastrophe. This archive is swollen with depictions of the disastrous tropics, and with the number of disasters rapidly multiplying, it only promises to expand. From afar, depictions include satellite imagery of geophysical hazards themselves, such as hurricanes, volcanoes, earthquakes, tsunamis, wildfires, floods, droughts, avalanches, and landslides, as well as close-up depictions of their on-the-ground consequences for landscapes, nature, and people.

PUERTO RICANS AS CLIMATE REFUGEES?

"Puerto Ricans Could Be Newest US Climate Refugees," read a headline on a September 28, 2017, article published in the authoritatively named *Scientific American* magazine.[16] "America's Climate Refugees Have Been Abandoned by Trump," read another in *Mother Jones* magazine on October 17, 2017. On December 3, 2017, the Morningside Center for Teaching Social Responsibility filed, under "Current Issues," a lesson for children titled "Cultivating Compassion for Puerto Rico's 'Climate Refugees.'"[17] An article published on February 19, 2018, in *Yes!* magazine was titled "Climate Refugees in Florida Could Change the Politics There for Generations."[18] On May 7, 2018, *Teen Vogue* published a poignant firsthand account titled "Hurricane María Made Me a Climate Change Refugee."[19] These headlines are integral to the greater ideological construction classifying Puerto Rican migrants post-María as climate refugees in the zeitgeist.

These claims open the field to various inquiries: Who are climate refugees? What makes a population qualify as such? What conventions guide such a definition? And what does the application of this label mean for the labeled population? In the 1970s the term *environmental refugee* emerged to describe people who were forced to migrate as a result of environmental deterioration or catastrophic conditions in their homelands.[20] In 1985 a formal definition of *environmental refugee* entered the public discourse when Essam El-Hinnawi, working with the UN Environment Programme, defined them as "those people who have been forced to leave their traditional *habitat*, temporarily or permanently, because of marked environmental disruption (natural or triggered by people) that jeopardized their existence and/ or seriously affected the quality of their life [emphasis added]."[21]

The notion of traditional habitat is problematic for at least two reasons. When used to refer to islands, it glosses over the fact that migration to and from islands has historically followed the ebb and flow of natural climatic hazards. In the case of Puerto Rico's history since the European conquest, migration has been closely tied to economic conditions. Movement of populations in and out of islands has been part and parcel of Caribbean habitation since prehistoric times.[22] Perhaps it is time to revisit the idea that people born on islands, such as Puerto Rico, are meant to remain there for the duration of their lives. Conceptualizing Indigenous and Black populations from the Global South as people who "naturally" belong to their ecosystem, like plant species, frames their mobility and migration as a pathological condition of uprootedness, contrasted with wealthy cosmopolitan Westerners who are highly mobile, global citizens.[23] Though there is no clear difference between the terms *environmental* and *climate refugee*, it appears as if *environmental refugee* morphed into *climate refugee* as a way to more directly index the role climate change plays in causing population displacement and migration.[24]

Thus, the media narrative classifying Puerto Ricans as climate refugees ignores two important issues. First, the 1951 Refugee Convention does not recognize "environmental" or "climate" refugees. These categories of displaced people as deserving protection emerged later.[25] Under the 1951 Convention, refugees are defined only as those who fear being persecuted "on the grounds related to race, religion, nationality or membership of a particular social group or political opinion, and are unable or unwilling, for fear of persecution, to seek protection from their home countries."[26] Second, because Puerto Ricans are US citizens, they are not considered refugees since they did not cross international borders. Instead, according to the 1998 UN Guiding Principle of Internal Displacement, they are "internally displaced persons" (IDPs). Because the storm and its disastrous consequences are understood to produce short-duration hardships, Puerto Ricans do not face any of the three impediments—legal, factual, or humanitarian—for returning to their homeland, as defined in a 2012 UN report.[27]

Although the scholarship on Puerto Rican migration is plentiful, it has not sufficiently emphasized that Puerto Ricans have long been economic and climate migrants to the continental United States. In the twentieth century, increasing modernization and eventual industrialization led to the total collapse of agro-industrial jobs on the island. In turn, thousands of individuals and families migrated north. On the island, the feminized industrial sector was never enough to lower the ranks of the unemployed, and chronic joblessness among working-age men and women coupled with episodic hurricanes and droughts also contributed to people's decision to migrate. Moreover, episodic emigration and return have been a characteristic livelihood adaptation for many Puerto Ricans.[28] In the twenty-first century, the island's debt crisis and ensuing austerity policies and politics, coupled with intensifying climatic hazards, contamination, and pollution, have led more people to leave the island.

The media's construction of Puerto Ricans as climate refugees served several ends. It underscored the catastrophic nature of the hurricane, as well as its traumatic consequences. It is widely believed that the conditions leading to refugee status always produce trauma in the individuals or groups who experience it.[29] The popular images and cultural characterizations of Puerto Rican climate refugees made them a "target population" whose "behavior and well-being" would be closely examined by the public as well as by policy makers. "Target population" narratives and visual characterizations are evaluative, thereby allowing for the public at large and, more significantly, policy makers to make assertions about a group's deservingness.[30] What is useful about the target population framework is that it helps explain how the needs of populations are co-constructed. The media constructions that use this framework are often built on preexisting racial, classed, and gendered assessments. They thus re-inscribe some groups as more advantaged than others, and more importantly, the policies they endorse often reinforce long-standing beliefs about the specific population at the center of public discourse. In the case of post-María Puerto Ricans, victimizing tropes such as helplessness and dependency (i.e., culture of poverty) reemerged in the cultural and political arenas.

The construction of Puerto Ricans as climate refugees has thrust them into the center of the discourse on climate change crisis. This has had several effects on the population itself as its plight is used as evidence of the climate crisis, positioning them as victims mired in inequitable power relations.[31] In terms of the design of public policies, the historically negative evaluation of Puerto Ricans as dependent on the state has also constructed them as burdens on the state. The recent depiction of the "Puerto Rican exodus," as arriving to inflict an even greater burden on the state's social services, has served to promote racist fears in stateside locations where Puerto Ricans have sought refuge. Certainly, an "exodus" of Puerto Rican climate refugees draws on negative, racialized cultural referents of refugees as impoverished, needy, and destitute. Additionally, the current US political climate, in which moral panics have been stoked about security at the southern border, has increasingly conflated and strengthened racist associations between "refugees" and the dangers to American society supposedly posed by "illegal immigrants."

In the case of post-María Puerto Rican climate migrants, the negative valuation of their needs has been exemplified in FEMA's abrupt ending of the Transitional Sheltering Assistance program, which offers lodging assistance to people who lost their homes. Though the initial response to the crisis appeared robust in "friendly states" (such as Connecticut, for instance), within a short span of time the humanitarian impulse had diminished, and many climate migrants were left to their fate.[32] In cities such as Orlando, Florida, which has in the last two decades turned into a Puerto Rican ethnic enclave and is in the throes of a housing shortage, climate migrants have been unable to find affordable housing. Speaking from Orlando a year after Hurricane María, seventy-year-old Benjamin Muñoz explained, "We have come here and have been treated badly, they think we are bad people. Sometimes I say, 'do we have something contagious?' We have been abused we have suffered discrimination. . . . It bothers me when I hear that the crisis has been solved, that we have been given apartments, there is a large group of people who still don't have any place to live."[33]

Clearly, the suffering that began with Hurricane María's landfall in Puerto Rico has continued to reverberate across space and time; the storm itself marks only the beginning of the crucible experienced by those individuals displaced by climate. The increasing ferocity of climate change is only expected to increase the suffering of individuals around the world. If the goal is to prevent death and suffering, governments must implement policies to ameliorate the social factors, such as inequality and careless development, that turn a climate hazard into an actual disaster. Though Puerto Rico and Puerto Ricans have long been imaged within the visual regime of the disastrous tropics, these latest depictions thrust them into a historically unacknowledged or, for some, seemingly impossible category: the Puerto Rican "climate refugee." This is a category of migrants that is on the rise worldwide, and while the case of Puerto Rico and its people is exemplary, it is hardly exceptional.[34] This means the next debt crisis or environmental disaster may well affect vulnerable communities near you.

1. Derek Walcott, "The Antilles: Fragment of Epic Memory," Nobel Lecture, December 7, 1992, https://www.nobelprize.org/prizes/literature/1992/walcott/lecture/.

2. Danica Coto, "Puerto Rico OKs Airport Privatization amid Protests," *USA Today*, March 3, 2013, https://www.usatoday.com/story/todayinthesky/2013/03/01/puerto-rico-airport-privatization-deal-lifts-off/1956407/; John Kosman, "Canada Just Bought Half of Puerto Rico's Main Airport," *New York Post*, March 22, 2017, https://nypost.com/2017/03/22/canada-just-bought-half-of-puerto-ricos-main-airport/; Danica Coto, "Puerto Rico Airport to Unveil 200M in Upgrades," *USA Today*, July 2, 2014, https://www.usatoday.com/story/todayinthesky/2014/07/02/puerto-ricos-san-juan-airport-to-unveil-200m-renovations/11943569/.

3. David Begnaud, "After Hurricane María, Puerto Rico Airport Looks Like a Shelter," *CBS This Morning*, September 25, 2017, https://www.youtube.com/watch?v=dBqoNpOkqXE; Pablo Venes, "Puerto Rico's Biggest Airport Becomes a Refugee Camp," *Daily Beast*, September 27, 2017, https://www.thedailybeast.com/puerto-ricos-biggest-airport-becomes-a-refugee-camp.

4. Venes, "Puerto Rico's Biggest Airport."

5. Patrick Gillespie, "'I Don't Feel Safe Here': A Night of Desperation in San Juan's Sweltering Airport," CNNMoney, September 27, 2017, https://money.cnn.com/2017/09/27/news/puerto-rico-san-juan-airport/index.html.

6. Venes, "Puerto Rico's Biggest Airport."

7. David Koenig and Danica Coto, "Puerto Rico, with Almost No Electricity, Endures Stifling Heat," *Seattle Times*, September 25, 2017, https://www.seattletimes.com/business/puerto-rico-is-in-the-dark-in-wake-of-hurricane-maria/.

8. Gillespie, "'I Don't Feel Safe Here.'"

9. Richard Fausset and Alan Blinder, "At Puerto Rico's Main Airport, Heavy Hearts and Long Waits," *New York Times*, September 26, 2017, https://www.nytimes.com/2017/09/26/us/puerto-rico-airport-maria.html.

10. Oren Dorell, "Who Will Rebuild Puerto Rico as Young Professionals Leave Island after Hurricane María?," *USA Today*, October 12, 2017, https://www.usatoday.com/story/news/nation/2017/10/12/puerto-rico-young-professionals-leaving-hurricane-Maria/754753001/.

11. Jack Healy and Luis Ferré-Sadurní, "For Many on Puerto Rico, the Most Coveted Item Is a Plane Ticket Out," *New York Times*, October 5, 2017, https://www.nytimes.com/2017/10/05/us/puerto-rico-exodus-Maria-florida.html. See also Frances Negrón-Muntaner, "The Emptying Island: Puerto Rican Expulsion in Post-María Time," *E-misférica*, no. 14 (2018): http://beta.hemisphericinstitute.org/en/emisferica-14-1-expulsion/14-1-essays/the-emptying-island-puerto-rican-expulsion-in-post-Maria -time.html/

12. Hilda Lloréns, *Imaging the Great Puerto Rican Family: Framing Nation, Race and Gender during the American Century* (Lanham, MD: Lexington Books, 2014). See also Allan Sekula, "The Body and the Archive," *October* 39 (1986): 3–64; Walcott, "The Antilles," Nobel Lecture; Sylvia Wynter,

"Unsettling the Coloniality of Being/Power/Truth/Freedom: Towards the Human, after Man, Its Overrepresentation—an Argument," *CR: The New Centennial Review* 3, no.3 (2003): 257–337.

13. Hilda Lloréns, "Imaging Disaster: Puerto Rico through the Eye of Hurricane María," *Transforming Anthropology* 26, no. 2 (2018): 136–56.

14. Lloréns, "Imaging Disaster," Transforming Anthropology, 136–56; Luis Ferré-Sadurní, Frances Robles, and Lizette Alvarez, "'This Is Like a War': A Scramble to Care for Puerto Rico's Sick and Injured," *New York Times*, September 26, 2017, https://www.nytimes.com/2017/09/26/us/puerto -rico-hurricane-healthcare-hospitals.html; Amy Aubert, "Fear after María as Dialysis Patients in Puerto Rico in 'Desperate Need' of Help," ABC7, September 29, 2017, https://wjla.com/news/nation-world/fear-after-maria -as-dialysis-patients-puerto-rico-in-desperate-need-of-help.

15. Union of Concerned Scientists, "Each Country's Share of CO2 Emissions," October 11, 2018, https://www.ucsusa.org/global-warming/science-and -impacts/science/each-countrys-share-of-co2.html.

16. Daniel Cusick and Adam Aton, "Puerto Ricans Could Be Newest US Climate Refugees," *Scientific American*, September 28, 2017, https://www .scientificamerican.com/article/puerto-ricans-could-be-newest-u-s-climate -refugees/.

17. Maireke van Woerkom, "Cultivating Compassion for Puerto Rico's 'Climate Refugees,'" Morningside Center for Teaching Social Responsibility, December 3, 2017, https://www.morningsidecenter.org/teachable-moment/lessons /cultivating-compassion-puerto-ricos-climate-refugees.

18. Adam Lynch, "Climate Refugees in Florida Could Change the Politics There for Generations," *Yes!*, February 19, 2018, https://www.yesmagazine.org /planet/climate-refugees-in-florida-could-change-the-politics-there-for -generations-20180219.

19. Agnes M. Torres Rivera, "Hurricane María Made Me a Climate Change Refugee," *Teen Vogue*, May 7, 2018, https://www.teenvogue.com/story /hurricane-Maria -made-me-a-climate-change-refugee.

20. Norman Myers, *Environmental Exodus: An Emergent Crisis in the Global Arena* (Washington, DC: Climate Institute, 1995); Norman Myers, "Environmental Refugees: A Growing Phenomenon of the Twenty-First Century," *Philosophical Transactions* B 357, no. 1420 (2002): 609–13.

21. Joanna Apap, "The Concept of 'Climate Refugee': Toward a Possible Definition," *European Parliamentary Research Service*, June 19, 2018, http://www .europarl.europa.eu/thinktank/en/document.html?reference=EPRS_BRI.

22. William F. Keegan and Corine L. Hofman, *The Caribbean before Columbus* (New York: Oxford University Press, 2017); Irving Rouse, *The Tainos: The Rise and Decline of the People Who Greeted Columbus* (New Haven, CT: Yale University Press, 1992).

23. Liisa Malkki, "National Geographic: The Rooting of Peoples and the Territorialization of National Identity among Scholars and Refugees," *Cultural Anthropology* 7, no. 1 (1992): 24–44; Cecilia Tacoli, "Crisis or Adaptation?

Migration and Climate Change in a Context of High Mobility," *Environment and Urbanization* 21 (2009): 513–25.

24. Apap, "Concept of 'Climate Refugee.'"

25. Carol Farbotko and Heather Lazrus, "The First Climate Refugees? Contesting Global Narratives of Climate Change in Tuvalu," *Global Environmental Change* 22 (2012): 382–90.

26. Apap, "Concept of 'Climate Refugee.'" It is striking—though not surprising, given the history of European colonialism and imperialism—that the 1951 convention altogether dismisses the central roles played by land, water, their territorial ownership, and their habitability (i.e., whether a place is suitable for people to make good lives) in originating many of the so-called political, rather than environmental, conflicts that give rise to refugee crises in the first place.

27. Walter Kälim and Nina Schrepfer, *Protecting People Crossing Borders in the Context of Climate Change* (Geneva: Division of International Protection, United Nations High Commissioner for Refugees, 2012), https://www.unhcr.org/4f33f1729.pdf.

28. David Griffith and Manuel Valdés-Pizzini, *Fishers at Work, Workers at Sea: A Puerto Rican Journey through Labor and Refuge* (Philadelphia: Temple University Press, 2002); Jorge Duany, *The Puerto Rican Nation on the Move: Identities on the Island and in the United States* (Chapel Hill: University of North Carolina Press, 2002).

29. Art Hansen and Anthony Oliver-Smith, *Involuntary Migration and Resettlement: The Problems of Response and Dislocated People* (Boulder, CO: Westview, 1982); Malkki, "National Geographic," 24–44; Dermont Ryan, Barbara Dooley, and Ciarán Benson, "Theoretical Perspectives on Post-Migration and Well-Being among Refugees: Towards a Resource-Based Model," *Journal of Refugee Studies* 21, no. 1 (2008): 1–18; Anthony Oliver-Smith, "Climate Change and Population Displacement: Disasters and Diasporas in the Twenty-First Century," in *Anthropology of Climate Change: From Encounters to Actions*, ed. Susan A. Crate and Mark Nuttall (Walnut Creek, CA: Left Coast, 2009), 116–38; Celia McMichael, Jon Barnett, and Anthony J. Michael, "An Ill Wind? Climate Change, Migration and Change," *Environmental Health Perspectives*, 120, no. 5 (2012): 646–654. https://doi.org/10.1289/ehp.1104375.

30. Anne Schneider and Helen Ingram, "Social Construction of Target Populations: Implications for Politics and Policy," *American Political Science Review* 87, no. 2 (1993): 334.

31. Carol Farbotko and Heather Lazrus, "The First Climate Refugees? Contesting Global Narratives of Climate Change in Tuvalu," *Global Environmental Change* 22 (2012): 382–90; Michael T. Bravo, "Voices from the Sea Ice: The Reception of Climate Impact Narratives," *Journal of Historical Geography* 35 (2009): 256–78; Geraldine Terry, "No Climate Justice without Gender Justice: An Overview of the Issues," *Gender and Development* 17, no. 1 (2009): 5–18.

32. Frances Robles, "Government Can Stop Paying to House Puerto Rico Hurricane Victims, Judge Rules," *New York Times*, August 30, 2018, https://

www.nytimes.com/2018/08/30/us/puerto-rico-fema-housing.html; Sarah
Ruiz-Grossman, "Displaced Puerto Ricans Face Dire Situations as FEMA
Housing Aid Nears Its End," August 24, 2018, HuffPost, https://www
.huffingtonpost.com/entry/fema-housing-aid-hotels-puerto-rico-hurricane
-maria-survivors_us_5b7f2608e4b0729515115850.

33. Kate Santich and Carlos Vásquez Otero, "Puerto Rican Evacuees in Central
Florida Seek Stability in a New Life 1 Year After María," *Orlando Sentinel*,
September 21, 2018, https://www.orlandosentinel.com/news/puerto-rico
-hurricane-recovery/os-hurricane-maria-one-year-later-central-florida
-20180921-story.html.

34. Yarimar Bonilla, *Non-sovereign Futures: French Caribbean Politics in the
Wake of Disenchantment* (Chicago: University of Chicago Press, 2015);
Aarón Gamaliel-Ramos, *Islas migajas: Los países no dependientes del Caribe
contemporáneo* (San Juan: Travesier and Leduc, 2016).

ACCOUNTABILITY AND REPRESENTATION

Photographic Coverage after the Disaster

Erika P. Rodríguez

When Hurricane María hit that Wednesday morning, what was to come in the following days, months, and year was unimaginable. It was my first storm season working as a photographer, and I was to learn as things unfolded. Like others in Puerto Rico, I suffered damages to my home. I worried about my family and wondered where I was going to find food. I stood in long lines for gas and cash, was eaten alive by mosquitoes every night, and lived in the darkness of the power outage. There was no way to photograph it from a distance, merely as an observer, because I was part of it. This was also my story.

Since 2012, I have been documenting my homeland with what began as a personal project and has developed into my largest body of work. "The Oldest Colony" is a meditation on Puerto Rican identity through the framework of our colonial relationship with the United States. I started photographing as a process to understand my own identity and experience as a Puerto Rican. It began as a personal matter when I took the camera with me and photographed my family, my friends, and the places that were quotidian for me, but it later grew into a documentation of festivals, political events, protests, and daily life, as well as assignment work. In the pictures, I am always in search of that underlying layer of mild tension that expresses the political limbo and identity crisis we exist in.

Our collective memory is fragmented, which is something that I dwell on and that influences my creative process. Photography has been a means to fill in the gaps of my fractured knowledge and education about Puerto Rico and its history. By creating a visual language and representation of our times, I want to leave a body of work that in fifty or one hundred years can give a glimpse of our reality, culture, joys, and hardships. I often ask myself, "How are we going to look back at this time, and what are the images that are going to represent what we're living through right now?"

Ours is a story of resilience, given that we carry over five hundred years of colonization. Puerto Rico's political condition as an unincorporated territory of the United States, of existing under the discretion of Congress, of not being a country—of remaining a colony—has denied us the right of self-determination, autonomy, and sovereignty. "The Oldest Colony" is not just something I talk about, it's something I live. It's that struggle and that constant internal search to define who we are and, in the process, who I am.

After covering the damage left by Hurricane Irma in the Virgin Islands for the *New York Times*, a new storm was announced. On a regular Sunday morning, I was sitting at a café in Santurce when the news alert came in: a category 1 storm was expected to strengthen and hit the island directly. We had two days to prepare. On September 20, 2017, Hurricane María became the worst natural disaster to strike Puerto Rico in modern times. The storm knocked out power and communications, closed schools and hospitals, and created a mass exodus. It took a while to understand the depths of the disaster. When arriving to a community in the first weeks and months after the storm, I was often asked if I was an officer from the Federal Emergency Management Agency (FEMA). People desperately needed help. People often said, "Ustedes son los primeros que llegan aquí" (You're the first ones to get here). All I had was my camera.

As many as 2,975 residents died as a result of the storm, according to a report by George Washington University. The official death toll is derived from the estimate, but no actual numbers are expected to be determined. Until everyone has a roof over their heads and until all the dead have been named, no recovery can be declared.

Our political limbo permeates everyday life. We have a rich culture, but we don't know who we are as a nation, or nonnation. Throughout history the images that represent the commonwealth have been too often filled with stereotypes that minimize and hinder our interactions.

Helpless people. Unfit. Unable. Welfare Island. Paradise. Entertainment. Escape. Coconuts and piña coladas. The place Americans can travel to without a passport.

Through the subtlety of daily life, the chaos of politics, the hardship of the economic crisis, and the hurricane's recovery our narrative emerges—with its celebrations and contradictions. There were images I didn't make in the coverage of María. In the aftermath, I felt the responsibility to create images that represented people with dignity, beyond the loudness of the disaster. It was not about bending the truth or hiding it, but about understanding a situation and deciding what is the most responsible way to document people when they are most vulnerable.

Photojournalists are there to capture things as they are happening, but we have a responsibility to respectfully represent the communities that are allowing us to be there with our cameras, especially in situations of pain and trauma. There is a fine line between documentation and exploitation. I tend to think before pressing the shutter. I work slowly. I try to avoid reproducing an image of helplessness, a "look at the poor brown people in the Caribbean, let's go help them and bring them everything—because they don't know how and can't help themselves." The people in my photographs exist in my common spaces; they are people I might see again. It's my neighbor, it's the person putting gas in the pump next to me, it's the person in front of the line at the supermarket, in a festival, at a friend's house. There's a different relationship when you work in a place where you're accountable for how you represent it.

After the hurricane many birthdays on the island were celebrated in communities without electricity, drinkable water, or access to fresh food. In December, Christmas came as thousands remained in the dark or in shelters. It might have felt as if there were not a lot of reasons to celebrate, but fighting for our right to feel joy became for us not only a thing to do but also a survival technique. A means of saying we are here, we are struggling, we are together—we are not in our best conditions, but we are unified—and our strength, our reasons to be alive, are taking place right now, right here, in these celebrations.

The perseverance and courage of those who remain, even as they struggle to rebuild and continue their lives, is where light leaks through. It is in quotidian spaces, in the most simple acts of organizing, of solidarity and building community, that resilience lives.

Puerto Rico governor Ricardo Rosselló onstage with his family and other politicians from his New Progressive Party, celebrating the birth of José Celso Barbosa in Bayamón, Puerto Rico, on August 7, 2016. Barbosa is considered the father of the island's pro-statehood ideology.

A young couple listens to a speaker during a march for Puerto Rican independence in Hato Rey, San Juan, Puerto Rico, on June 11, 2017. The American territory held its fifth nonbinding referendum to choose what political relationship it wants with the United States. Amid the island's fiscal crisis, the event cost the local government $7 million.

Veteran Miguel Quiñones poses for a portrait in his home at Barrio Bubao in Utuado, on October 25, 2017.

Police officers walk through a cloud of tear gas in Hato Rey, San Juan, Puerto Rico, during the National Strike on May 1, 2018. For International Workers Day, people took to the streets in protest of the austerity measures implemented by the Fiscal Control Board and the local government.

Moráima Fuster serves ice cream to a customer at Heladería Lares, in front of Revolution Square in Lares, Puerto Rico, on June 5, 2017. The ice cream store is popular for its more than sixty unique flavors, including avocado and rice and beans.

Fourth grader Camila solves a problem during math class at the Hiram González Elementary School in Bayamón, Puerto Rico, on May 9, 2017. The public school admitted almost 200 more students from a neighboring elementary school that was closed as part of a consolidation plan by Puerto Rico's Department of Education.

People say goodbye to a loved one at the Luis Muñoz Marín International Airport in Carolina, Puerto Rico, on January 21, 2018. After Hurricane María ravaged the island on September 20, 2017, over one hundred thousand people are estimated to have left the island for the continental United States.

Eliezer Román cuts the weeds on a hillside for a new coffee crop at Hacienda Lealtad in Lares, Puerto Rico, on June 5, 2017. The owner of the plantation, Edwin Soto, has had to diversify his business to counter the losses in coffee production. Recently he opened a coffee shop and plans to open a hotel in the colonial-era estate of the Hacienda Lealtad.

An abandoned building that collapsed from the hurricane in Puerta de Tierra, San Juan, Puerto Rico, on September 21, 2017.

An aisle for frozen goods remains closed as the power generator of Ralph's Food Warehouse in Humacao, Puerto Rico, is not powerful enough to keep the freezers running, on October 23, 2017. A month after Hurricane María ravaged the island, food distribution was still unstable, leaving many shelves empty and bottled water scarce.

Jeremy Arce, 22, cleans parts of the ceiling that fell at the coffin showroom in Funeraría J. Oliver in Ponce, Puerto Rico, on November 8, 2017. The coffins have been covered in plastic to protect them from roof leaks. The funeral home said their services had tripled since Hurricane Maria devastated the island over a month earlier.

People look at the destruction left by Hurricane María hours after the storm passed the island on September 20, 2017, in Guaynabo, Puerto Rico. The storm landed with sustained winds of 155 miles per hour.

A woman poses for a portrait in her home in San Juan, Puerto Rico, on August 4, 2018. Almost eleven months after the storm destroyed her home, and affected many in her community, she still lived under plastic tarps.

According to the company, people rest as they wait in line throughout the night for the opportunity to purchase no more than two bags at Tropical Ice in Ponce, Puerto Rico, on September 28, 2017. Puerto Rico's power utility took eleven months to restore power to the entire island.

Julia Guzmán Serrano, 65, hangs a solar lamp in her living room on May 22, 2018, in Utuado, Puerto Rico. Eight months after the storm, her home remained without power. Her 67-year-old husband, William Reyes Torres, suffers from sleep apnea and could not power his breathing machine during the night. "I pray every night to wake up the next morning," said Reyes.

Carlos Torres draws in the darkness of the power outage while he spends time with neighbors outside at San Isidro community in Canóvanas, Puerto Rico, twelve days after Hurricane María on October 2, 2017.

From left: Lorraine Martínez, Tamary Díaz, Deliris Ortiz, and Keren Acevedo sing under a cell phone light during a parranda at the house of a coworker who still had no power in Guaynabo, Puerto Rico, on December 16, 2017. About half the island spent the holidays in the dark that year.

Leo Garay Bernal, 3, shows his tennis shoes to Waleska Semiday, as he looks for shoes his size at the impromptu memorial for people who died in the aftermath of Hurricane Maria on June 1, 2018, in San Juan, Puerto Rico. The event, called Project 4,645, was a collective initiative organized on social media as a reaction to the study by Harvard's T. H. Chan School of Public Health published earlier that week in the New England Journal of Medicine. The study estimated that 4,645 people died as a result of Hurricane María and its aftermath, in contrast to the government's official number, which remained at sixty-four direct deaths. By nighttime there were over eight hundred shoes, some of which had stories or names attached to them.

LIFTING THE VEIL

Portraiture as a Tool
for Bilateral Representation

Christopher Gregory

The idea that Hurricane María lifted a "veil" covering Puerto Rican realities was repeated in both local and international media. Perhaps the key driver of this conceptualization in the wake of the hurricane was the images circulating in outlets and social media. The role of the image in understanding the aftermath of disaster is undeniable in any situation, especially in the increasingly mediated modern world. But in Puerto Rico, something different happened. By lifting a veil, the images seemingly changed the way Puerto Ricans understood themselves; it revealed a truth that was somehow always there but unseen.

Behind the so-called veil was an island plagued with widespread poverty and decrepit infrastructure. But this reality had always been faced by a large part of the population on the island. For them the lifting of the veil revealed not their own poverty but the other side of the gaze: an "affluent" metropolis unwilling to accept the reality of failed industrialization.

My work and its methodology aim to disrupt this narrative of industrial morality. While the rural communities in many instances have, in fact, had many disadvantages, their material or circumstantial loss, in my opinion, should not be the main focus of documentation. In addition to the physical destruction, the psychological space inhabited by the residents of Puerto Rico was crucial to represent and understand. This psychological space is not only a result of the death and destruction caused by the hurricane but also the political reality revealed by the hurricane—a colonial power that has abandoned its subjects.

To avoid a post-Fordist gaze in my own documentation, I opted to use portraiture in addition to traditional vérité reportage. Portraiture in this context provides a collaborative representation of the subject. It exploits the evidentiary nature of photography to provide a document that is a dialogue, not a "capture." My methodology involved spending a long time with my subjects before photographing them, not only to understand their perspective on their situation but also to develop a relationship of trust for a meaningful exchange during the sitting.

Although the hurricane was a "news" event, I view the documentarian's role as one that must take into consideration the long historical context of the people and place at hand to get at the underlying systems that created the situation being documented. In the case of the storm, it is a colonial relationship that extends beyond present-day fiscal austerity and political aspirations. It includes imperial violence that has left hundreds of thousands of people across the Americas to fend for themselves.

My portraits, especially of rural communities, aim to connect the reader with their physical and psychological state but also to this historic reality that extends beyond the storm. For many rural communities, the abandonment came as no surprise. Before the storm, they were often isolated from even their own local government in the challenging terrain of the Cordillera Central.

With this body of work, it is my intention to illustrate this particular abandonment while exalting the radical and resilient independence that these communities have developed as a result of this violence. It is a spirit that transcends the political reality of colonialism and is ill-served by the imposition of a post-Fordist gaze.

The principal road of Punta Santiago in Humacao, Puerto Rico. Humacao was one of the municipalities on the southeast of the island where the eye of the storm made landfall and one of the last to have electricity restored.

A damaged home in Aibonito, a town in Puerto Rico's mountainous central region.

Javier Cabrera sifts through scraps at a makeshift junkyard set up in Aibonito after the hurricane. "I can't leave everything to the government," he said while looking inside discarded televisions for electronic parts, which he used to fix electrical generators.

In Punta Santiago, Humacao, a community organization hands out three bags of ice to residents. Many used it to keep food and medical supplies from spoiling during the long months without electricity.

Members of the National Guard talk to a boy in the Guavate neighborhood of Cayey. They offered him a chance to get in the driver's seat of a convoy vehicle. The guard as well as FEMA and the army have been criticized for inefficiently distributing meals and aid to the island.

A chainsaw and machete used by utility repair workers in Utuado, Puerto Rico.

Javier Jose Capó Alicea's home in Yabucoa, Puerto Rico, where Hurricane María made landfall.

An abandoned building in the town center of Utuado, which was flooded with up to five feet of water.

Mother and daughter Vivian Reyes and Gladys Rivera in their home in Utuado, in the room where they spent the hurricane until the roof was blown off. This photograph was taken on the first day they received official help.

Grave keeper Tulio Collazo Vega poses by the grave of three elderly sisters who were killed by a landslide in Utuado the day the hurricane hit Puerto Rico. According to Collazo, their ashes were buried almost a year later.

Magdalena Flores poses for a portrait in front of her home in Utuado. Magdalena's daughter died in the aftermath of the storm from complications of asthma.

San Juan mayor Carmen Yulín Cruz, who has led her own recovery efforts, separate from those of the central and federal government, out of a makeshift warehouse and shelter at the Roberto Clemente Coliseum. Her attacks on Trump launched her into the national spotlight during the hurricane.

Kaleb Acevedo poses for a portrait on the new bicycle he got after Hurricane María.

Nahielys Gonzales and Yamil Rodríguez in the town square of Utuado after school.

THE IMPORTANCE OF POLITICALLY ENGAGED ARTISTIC AND CURATORIAL PRACTICES IN THE AFTERMATH OF HURRICANE MARÍA

Marianne Ramírez-Aponte

Puerto Rico is in the midst of one of the most urgent crises in its history, and this crisis is not only political and financial but also social, cultural, and civic. The US Congress has enacted a law known as PROMESA to address Puerto Rico's debt crisis and has installed a Financial Oversight and Management Board with plenary powers over the budget and the resolution of the island's debt, thereby limiting the powers of the island's elected government. Those actions, along with Puerto Rico's virtual destruction by two hurricanes, have created a situation never before experienced in Puerto Rico or, in my estimation, anywhere else in the world.

This "exceptional situation" has been used to justify austerity measures and changes in public policy that have occurred very rapidly, one after another, affecting the quality of life of all residents. It has also triggered an unprecedented wave of migration that has had an impact on the Puerto Rican community in the United States, generating, especially after Hurricane María, a sense of connectedness and solidarity between the diaspora and the island. We have also seen an unprecedented expansion of the scope and frequency of public debates regarding the challenges facing Puerto Rico.

One of the positive effects of these concurrent events is that the Puerto Rico art scene has achieved greater visibility and works by Puerto Rican artists have been more widely disseminated. That is the result of a number of factors: events organized by academics; invitations for Puerto Rican artists to participate in exhibitions and residencies at museums and universities across the United States; and the publication of reviews and articles in the media and specialized journals and magazines, such as the *New York Times*, *ArtForum*, *Art in America*, and *Hyperallergic*.

Although this experience is not necessarily different from that of other artists in countries that have faced similar situations, I think it's important to point out that, historically, the ambivalent place Puerto Rico holds as a country in the community of nations is a factor that has contributed to keeping our artistic production from enjoying a more prominent international profile. That is, colonialism should be understood not only as a question of political governance but also as an eminently cultural problem. In the case of Puerto Rico, culture remains the most important basis for defining our national identity, and assumptions should not be made about the assimilation of the island's culture into that of the metropolis. I think it's necessary to make that distinction because the struggle for Puerto Rico's cultural and political independence is at the root of, and the objective of, a large part of the Puerto Rican art produced both on and off the island.

This is also true for the art created in the aftermath of the hurricanes that struck the island at the end of 2017. With the island's true situation revealed as never before, art—whose remarkable capacity for interpretation makes it a space of democratic participation—has been fundamental in creating a counternarrative, outside the ambit of officialdom, that is essential for understanding Puerto Rico after María. Artists have responded with images that not only capture the destruction caused by these meteorological events but also make visible circumstances, conditions, racial and economic legacies, and communities that have long been invisible or silent.

It is my belief—as I will discuss below regarding the work of the Museo de Arte Contemporáneo de Puerto Rico (MACPR)—that art museums, like artists and their work, take part in the task of "revealing" social realities, because they are sites of ideological power and of potential social agency and change. Learning to understand the meaning of images in a hypervisual world is a political act. In this respect, museums play an important role as formative spaces that provide education in visual literacy and critical-thinking skills as well as contributing to the dissemination of artworks through their contact with local and international audiences.

~*~

Artistic responses to Hurricane María thus took on a variety of forms. Every artistic discipline has been represented, including transdisciplinary collaborations, urban art, comics, internet art, interventions in the public space, performance, and community-based projects aimed at collective and individual recovery. These works have included collaborations and exchanges between artists who live on the island and those who live in other places, and the circular migration that has tended to characterize the practice of visual artists and that has intensified after the hurricanes. This circular migration takes place partly because professional opportunities have opened up, but especially because many artists have an interest in rethinking, from the perspective of our own identity, a new nation-building project.

Central to the work of these creators are issues associated with the slow and ineffective handling of the disaster by local and federal authorities; a colonial system in extremis and government corruption; the financial crisis and poverty; the fragmentation of families caused by migration; the transformation of the landscape and the potential loss of our cultural patrimony; climate change, dependency on fossil fuels, and environmental justice; and the disarticulation of the public patrimony, including an education system threatened by dismantling, among so many others.

As instances of artists' response to Puerto Rico's "new reality," Felipe Cuchí's *Ay, Carmela tu amor me estriñe!* (Oh, Carmela, your love constipates me!, 2017) and Migdalia Umpierre's series *Las cosas de María* (María's things, 2017) employed humor and the visual language of caricature to comment critically on the "new normal" and the deterioration of the residents' quality of life as they go without the provision of such basic services as electricity, potable water, and communications.[1]

Las cosas de María *("Maria's Things")*
by Migdalia Umpierre, 2017.

These works also present the implications to the island's citizens of nutritional insecurity in a place that imports almost 90 percent of its food and depends on imported fossil fuels for every aspect of daily life. They also denounce the government's shortcomings in managing the crisis and the delay in the island's reconstruction, with dreadful consequences for the economy and for people's lives. As evident for example in Carlos Dávila Rinaldi's *New Reality Social* (2018), the military presence and consequent state of siege speaks eloquently to the general sense of insecurity on the island, the desperate conditions faced by its citizens, and the interminable lines and other daily hardships met with in order to gain access to articles of basic need, leaving most citizens little time to engage in any political participation. All this points to the importance of seeing art—as it engages, informs, and activates people—as a vehicle for resilience and resistance in Puerto Rico.

New Reality Social by Carlos Dávila Rinaldi, 2018.

Other artists directly tackled how the storm transformed the surrounding landscape. For example, Yiyo Tirado's floor installation *Skyviews* (2017) presents aerial views of homes whose roofs were blown off by the hurricane and have been replaced by blue tarps.[2] The work presents forms whose repetition generates a sense of island residents' widespread experience of rooflessness and fosters an awareness of the island's poverty. The blue tarps distributed by FEMA are scars whose color and rigid geometry disfigure the landscape and point to Puerto Rico's dependency on the US government.

Sarabel Santos, on the other hand, takes a different approach to landscape in her series of floor installations titled *Groundscapes Displaced* (2017–18).[3] This series is based on the artist's experience and memory of the landscape before and after the hurricane, and it stems in part from her work photographing the devastation in the towns of Vega Alta, Vega Baja, and Bayamón. The varied compositions that make up the series are presented as fragmented grids of photographs that Santos configures at will. Their temporary installation in localities and landscapes of cities in the United States, when Santos temporarily stays there, generates a sense of dislocation among viewers who don't associate images of destruction with their clean, temperate surroundings. The documentation of the work where the installation has been shown corresponds to the artist's own displacement, her passage through these cities where she was occasionally stranded, and to the impossibility of carrying the landscape itself on her back.

To narrate her experiences after María, Rosaura Rodríguez, whose production involves recording daily life from both a social and a landscape perspective, created illustrated stories in ink and watercolor.[4] These stories, recently published in the graphic novel *Temporada* ([Hurricane] Season, 2018), document the various states of emotion experienced both individually and collectively as the days dragged on: restlessness, frustration, grief... In the case of Rodríguez, the transformations in the landscape had an impact on all aspects of her life as a woman who works on, and lives off, the land; as a botanical artist who acquires some of her pigments and materials from the earth; and as a teacher of art and ecology who, at her farm in Jayuya and in her project Camp Tabonuco, gives workshops to share the knowledge obtained with other artists and members of the community. Her work represents a successful model of self-governance, sustainability, and fair, healthy recovery for communities in a country without sovereignty.

Cover of Rosaura Rodríguez's graphic novel *Temporada*.

Thanks to its immediacy, its location in the everyday environment, urban art also contributes to raising awareness and activating citizens. Urban artists, using their murals painted in Puerto Rico and many cities in the United States, generate activism through images that denounce injustices and strengthen community ties and the island's bonds with the Puerto Rican diaspora. One example of this is the mural *MAY DAY 2018, Puerto Rico* by the collective Morivíví and the artist Sandra Antongiorgi, who lives in Chicago.[5] The mural was done as part of the Borinken Me Llama Public Art Series Festival in that city. It incorporates a quotation from the Puerto Rican abolitionist and *independentista* Segundo Ruiz Belvis: "There is no intermediate state between slavery and freedom." The image refers to the confrontations between the state security forces and the working class during the protest march held May 1, 2018, in San Juan.

MAY DAY 2018, Puerto Rico by Colectivo Moriviví, Chicago, Illinois

Here, the theme of slavery points to our colonial condition and the impoverishment of the working class as a result of the austerity measures imposed by the Financial Oversight and Management Board. The confrontation arises from the unequal distribution of wealth in the global capitalist system, which produces outsize wealth for owners and insecurity for workers. The iconography includes the shields carried by the workers that featured the Puerto Rican flag rendered in black and white, as if signifying mourning; the shields openly referred to the Puerto Rican flag painted on the door of an abandoned building in San Juan on July 4, 2016, by the collective Artistas Solidarixs y en Resistencia.[6] This image went viral and has been appropriated as a graphic element by many artists, becoming a commanding presence in the popular imagination.

In the same vein, other collectives composed of both local artists and artists in the diaspora, such as Agitarte, share that spirit of outrage and believe that art must be taken directly to communities.[7] This is the case with *End the Debt! Decolonize! Liberate! Scroll Project* (2018), a piece produced after Hurricane María in response to the United States' colonial policy toward Puerto Rico.[8] This is a portable object that is activated when it is unrolled by citizens who gather to do a shared "reading" of the current situation of Puerto Rico. The work, as its title indicates, includes images from the history of Puerto Rico and images of community activism that bear witness to the many struggles of the Puerto Rican people and its diaspora throughout more than a hundred years of colonial domination. The vignettes portray the landscape as a battleground, co-opting the traditional subjects of landscape in history for a history centered around protest and contestation.

As I have noted, condemnation of the colonial system is a recurrent concern of Puerto Rican artists. With the series *Portraits, from the Untitled Series on Puerto Rico Financial Oversight and Management Board Members, A Protest* (2018), artist Ricardo Hernández-Santiago joins the long tradition of Puerto Rican anti-colonialist art. In dialogue with the pictorial strategies of that tradition, Hernández-Santiago has created a somewhat grotesque portrayal of the Fiscal Control Board's members. The approach to the faces, which seem ready to break out of the picture plane, and the use of a gold background associated with wealth and power, emphasize the board's focus: that Puerto Rico make good on its obligations to its creditors, many of whom are private American investors. The series as a whole becomes an incisive portrait of the United States' current colonial policy, which is focused on collecting the debt without much regard for the social effects of the crisis.

Portrait of Natalie Jaresko, from the Untitled Series on Puerto Rico Financial Oversight and Management Board members; A protest, by Ricardo Hernández-Santiago, 2018.

With respect to the generalized sense of restlessness, unease, and frustration, nothing has caused more indignation and grief among the people of Puerto Rico than the loss of human life and the government's delay in recognizing the true number of deaths as a direct and indirect result of the hurricane and of the interruption of basic services.

Among the artistic projects addressing this issue, the one that received the most local and international media coverage was conceived by writer Rafael Acevedo, artist and critic Nelson Rivera, and writer and craftswoman Gloribel Delgado Esquilín. This group sent out a call via social media to the citizens of Puerto Rico to bring in shoes belonging to their loved ones who had died; the shoes were to be laid out on a plaza across the street from the capitol building in San Juan as a denunciation of government inefficiency in handling the crisis. (See image in the essay by Erika Rodríguez in this volume) The large-scale installation included almost three thousand pairs of shoes and stories that people shared about the victims. The act, scheduled for June 1, 2018, lasted forty-eight hours and was the first place where people had been able to meet for a time of collective mourning. The artists documented the act with photographs and note cards archiving each pair of shoes along with their stories; they intend to exhibit and publish a book with this material so that it will become a historical document of the hurricane's aftermath.

—◆

The multiple challenges that Puerto Rico faces at this moment have thus been addressed by artists with wide-ranging creativity. But empowering the space of art and its historical visibility at this difficult juncture should not be seen as the exclusive responsibility of artists. Socially engaged museums—with their ability to generate debate and enter into a dialogue with the reality that surrounds them, their commitment to the image and its analysis—can also be an important part of the recovery process. They can help bring to the collective mind images and a better understanding of the precariousness of Puerto Rico's history and the ambitions of its people.

I believe museums in Puerto Rico should acknowledge their political histories and adopt positions on contemporary issues and the Puerto Rican condition. Such is the case of the Museo de Arte Contemporáneo de Puerto Rico (MACPR). Our work, particularly during the last decade, has taken a two-pronged approach through socially conscious exhibitions and community outreach programs. In the post-María scenario, we have used the art in our collection as a means of protest and have fostered education and exchange as a means of shaping the way people reinterpret history—thus recalibrating power.

From that perspective, the exhibition *Entredichos*, which we organized as an emergency project and opened to the public in December 2017, revealed details of the social and political conditions of a particular moment in our history and proposed an approach to our culture from the standpoint of the tensions, contradictions, and complexities that define it.[9] That is why we chose the title *Entredichos*, a term that, in Spanish, suggests doubt as to the honor, veracity, or possibilities of a person or thing. In legal terms, *entredicho* is used in some jurisdictions to mean "injunction," a court-ordered halt to a potentially harmful or dangerous action. In this case, the curatorial intent was to make viewers stop, question the "givens," and, if necessary, condemn.

The exhibit was our response to the moment of grave collective danger we face in Puerto Rico today, a situation that has exacerbated the need for community organization and the need to build greater political and social power for Puerto Ricans. We also felt it was urgent to respond in our own voice(s) to the broad interest generated internationally by Hurricane María's destruction and our socioeconomic situation, and the many visions, both informed and uninformed, held on the island and abroad as to our history and the complex political relationship between Puerto Rico and the United States—a relationship that determines the socioeconomic structures that affect virtually every aspect of Puerto Ricans' lives.

The cultural struggles in Puerto Rico and the United States and our constant negotiation with history were evident across various pieces, including *Collective Ingenuity, or the Curse of the Parrot* (2010, revised 2012, 2014, 2017), by Elsa María Meléndez (viewable at https://vimeo.com/253547917). This installation, last revised by the artist after Hurricane María, on the one hand condemns the United States' interventionist policy toward Puerto Rico, manifested by the passage of the PROMESA law, and the imposition of the Financial Oversight and Management Board. On the other hand, the work pays homage to the student movement and its struggle against the dismantling of the University of Puerto Rico and public education on the island. In her latest updates, Meléndez alters the composition to include, on the wall that serves as a backdrop to the work, the form of Puerto Rico made up of countless faces. At the bottom of the map there are headless bodies, accompanied by dogs as symbols of the forces of state security. The debris on the floor at the center of the piece includes strips of cloth, shoes, necklaces, buttons, and pieces of wood pallets. These items evoke the barricades used by striking students and the chaos after the hurricane. Here we also find the hands of the dismembered bodies, positioned as in a sign of distress, to symbolize the anonymity of the three thousand deaths denied by the state. And as though that message were not clear enough, Meléndez underscores it by incorporating hashtags, such as #salvemoslaUPR (#letssavetheUPR), #noalaJunta (#nototheBoard), and #teodioMaría (#IhateyouMaría). The audio of a parrot that accompanies the piece becomes, then, a metaphor for a discourse endlessly repeated.

Collective Ingenuity, or the Curse of the Parrot by Elsa María Meléndez (2010, revised 2012, 2014, 2017).

The physical center of *Entredichos* was occupied by Pablo Delano's installation *The Museum of the Old Colony* (2015–17).[10] Its placement in a separate gallery within the MACPR reinforced the concept of a museum of history and anthropology within a museum of art. This work of conceptual installation art takes its name from a brand of soft drink that has been consumed in Puerto Rico since the 1950s, as a reminder that the island, an unincorporated territory of the United States, is considered the world's oldest colony.

This "museum" collection comprises reproductions of archival photographs of Puerto Rico, most of which were commissioned by agencies of the US government. The purpose of these commissions was to produce propaganda for a North American and worldwide audience portraying an idealized vision of Puerto Rico's progress under American rule. As a strategy to facilitate the viewer's interpretation of the images, Delano mutes his voice as an artist, allowing the viewer to share the critical space and to face the colonizer with no intermediation. Not only the images but also the accompanying captions reveal the colonizer in its full arrogance, racism, sexism, and misogyny.

Intermezzo: Ship Adrift by Garvin Sierra, 2017

At this moment Puerto Rico stands at a crossroads: on the one hand, a dead end; on the other, a dark and dangerous precipice raised by models of society now rendered obsolete; the road ahead is dark. In this spirit the work *Intermezzo: Ship Adrift* (2017), by Garvin Sierra, suggests that we Puerto Ricans are the twenty-first-century *balseros*. Trumpeted to the world in the 1950s as the showcase of democracy and a new model of progress, Puerto Rico today is a hallucinatory ship, filled with obsolete objects and unwieldy materials; it tries to propel itself by the pitiably weak force of a rusty old fan. The ship's sail, a faded Puerto Rican flag, points northward, but the ship doesn't move, because there is no wind. The artist's revised version of the piece after President Trump's visit in the immediate aftermath of Hurricane María takes the form of a buoy made of rolls of paper towels that blocks the ship's passage. It is a reminder of the president's gesture of contempt toward the Puerto Rican people at that moment and his continued disparagement of us ever since.

~◦

As I have noted, the MACPR's multipronged approach includes community-engagement programs so that the museum's cultural activity is projected outside its walls to the community. With this in mind, the museum implements a program we have called "The MAC in the Barrio: From Santurce to Puerto Rico," which uses the arts and culture as tools to contribute to cultural equity and social and urban transformation. Since 2014 the program has had a presence in several communities in the municipalities of San Juan, Guaynabo, and Cataño. The high percentage of people who live below the poverty level in these neighborhoods has made the MACPR's presence a pressing civic responsibility and has led to the decentralization of its services and the creation of programs that spill out of the metropolitan area, where Puerto Rico's largest cultural entities are located, to bring greater cultural activity to the entire island.

Among the objectives of "The MAC in the Barrio" is to critically interpret the historical, economic, political, and social fabric out of which our neighborhoods—our barrios—emerge and to make visible the distinct social communities that coexist in them. In the aftermath of María and throughout 2018, the museum carried out ten art projects in communities in San Juan, Cataño, and Guaynabo. The projects addressed such subjects as environmental justice, housing justice, gender violence, race, migration, cultural history, and community identity. The projects proposed and designed together with the community are promoted under the banner of inclusion, coexistence, and integration of the various social sectors, always watchful for conditions that might trigger undesired gentrification.

Two examples of these projects are Glorimar Marrero's *Juana(s) Matos* and Coco Valencia's *Lo onírico del Caño* (The dreamlike aspects of the Channel) (2018).[11] The first is a series of three short films about the contributions of Afro–Puerto Rican community leader Juana Matos and the reaction of residents of the neighborhood named after her in Cataño to the threats of displacement they have experienced throughout the years and especially since Hurricane María. The second is an installation of houses floating on the water of the Martín Peña Channel; on their walls are residents' statements about the physical and emotional loss they suffered during and after Hurricane María. *Lo onírico del Caño* is constructed out of the experiences of the residents of this settlement, a community that emerged in the mangrove swamps in this area some ninety years ago as part of the intramigrations that occurred on the island after Hurricane San Ciprian (1932) and during the Great Depression. Both projects, Marrero's and Valencia's, ask questions about how to respond to climate change in a fair and just way, a way that reflects the fact that it is the poorest among us who suffer the greatest impacts of environmental pollution and destruction.

Lo onírico del Caño (The dreamlike aspects of the Channel) by Coco Valencia, 2018

The museum's call to the public to learn about these truths and the broad dissemination of these and other projects generated and sponsored by the museum have inserted these issues into the collective consciousness. They stir the collective mind to social and environmental awareness and inspire citizens to rethink the status quo in order to arrive at a new ethical and civic order. The potential for critical thought must be stimulated so that new models of what the reconstruction and transformation of Puerto Rico should be may be envisioned. This must be undertaken while keeping in mind the importance of conserving our cultural patrimony without losing sight of the inequalities and impoverishment that any such models may generate or the destruction of our ecosystems caused by the exploitation of our natural resources.

In conclusion, the arts are intrinsically a social medium, both personal and public. They defend our humanity and help remove social barriers to give us all a better chance to progress. The work created by Puerto Rican artists in the context of our dire socioeconomic situation and in the aftermath of the hurricanes should be valued not only for its power to heal and prevent cultural erosion but also for being essential to amplifying the incomplete narrative of Puerto Rico's colonial crisis and ensuring the connectedness of the creative community with human ecosystems that need constant invigoration from new ways of interpreting the world and our island at this moment so crucial to our history and future.

1. Dianne Brás Feliciano, Emilia Quiñones Otal, and Mariel Quiñones Vélez, "The Impact of María on the Creative Community in Puerto Rico," February 6, 2018, https://transformaccpr.com/2018/02/06/el-impacto-de-Maria-en-la-comunidad-creativa-en-puerto-rico/; Puerto Rico Art News, "'Aftermath' Recent Work by Carlos Davila Rinaldo in Arte @ Plaza," April 8, 2018, https://www.puertoricoartnews.com/2018/04/aftermath-obra-reciente-de-carlos.html.

2. Artishock, "Yiyo Tirado: Caribbean Blues," *Artishock: Revista de arte* contemporaneo, http://artishockrevista.com/2017/12/19/yiyo-tirado-caribbean-blues/.

3. Sarabel Santos, "Groundscapes Displaced," http://www.sarabelsantos.com/groundspaces

4. Rosaura Rodríguez, "Temporada," https://www.instagram.com/p/BuZDcrwBhXF/

5. Participating artists in Moriviví include Raysa Raquel Rodríguez and Shanron "Chachi" González Colón.

6. 80 Grados, "The Flag Is Black, Puerto Rico Is Fighting," *80grados prensasinprisa,* https://www.80grados.net/la-bandera-esta-de-negro-puerto-rico-esta-en-pie-de-lucha/.

7. Participating artists and cultural workers in Agitarte include Jorge Díaz Ortiz, Estefanía Rivera, Crystal Clarity, Rafael Schragis, Emily Simons, Dey Hernández, Saulo Colón, and Osvaldo Budet.

8. "End the Debt! Decolonize! Liberate! Scroll Projects," *AgitArte,* http://agitarte.org/projects/end-the-debt-decolonize-liberate-scroll-project/.

9. Marianne Ramírez-Aponte, "Entredichos: New acquisitions of the permanent collection of MACPR," Museo de Arte Contemporaneo de Puerto Rico, http://mac-pr.org/entredichos-2018.html.

10. David Gonzalez, "A New Museum for an Old Colony, Puerto Rico," *The New York Times*, November 15, 2017, https://lens.blogs.nytimes.com/2017/11/15/a-new-museum-for-an-old-colony-puerto-rico/; "The Museum of the Old Colony," Hampshire College Art Gallery, https://sites.hampshire.edu/gallery/the-museum-of-the-old-colony/.

11. Mariela Fullana Acosta, "Glorimar Marrero goes through the story of Juana Matos," *El Nuevo Día,* November 11, 2018, https://www.elnuevodia.com/entretenimiento/cultura/nota/glorimarmarrerorecorrelahistoriadejuanamatos-2458869/; Museo de Arte Contemporaneo de Puerto Rico, "MAC in the neighborhood presents 'Lo Onirico del Cano,' a project by Coco Valencia," November 16, 2018, http://90grados.com/arte/mac-en-el-barrio-presenta-lo-onirico-del-cano-un-proyecto-de-coco-valencia/.

SI NO PUDIERA HACER ARTE, ME IBA

The Aesthetics of Disaster as Catharsis in Contemporary Puerto Rican Art[1]

Carlos Rivera Santana

> *Here, in this extremest danger of the will, art approaches, as a saving and healing enchantress; she alone is able to transform these nauseating reflections on the awfulness or absurdity of existence into representations wherewith it is possible to live: these are the representations of the sublime as the artistic subjugation of the awful.*
>
> —Friedrich Nietzsche[2]

Shortly after Hurricane María devastated Puerto Rico, the island's historic Museum of the Americas made a call for artists of all media to create works to "purge, purify and liberate their feelings" about their posthurricane experiences.[3] Over fifty artists contributed performance art, theater, poetry, music, art workshops, and more to the event, held December 10, 2017, entitled *Catarsis: Re/contruyendo después de María* (Catharsis: Re/constructing after María). This event crystallized the urgency of using art as a vehicle for (social) catharsis, a practice that continues to be used by individual artists, collectives, community organizations, art projects, and other art institutions on the island and abroad, through mural art, community paintings, art exhibitions, literature, music, and many other aesthetic expressions. The affective process of catharsis through art does not refer in this case to using art to overcome psychological disorders in individual therapy. Here, using art for social catharsis refers to an aesthetic process in which people can collectively express the complex or contradictory social, cultural, and political situations that confront them. They accomplish this through the successful transfiguration of complex or contradictory realities—such as the ones experienced by Puerto Ricans in the aftermath of Hurricane María—into another intelligible form or medium.[4]

Every cathartic activity requires a genre or narrative frame that facilitates the purge and liberation of a given affect, especially in the case of traumatic experiences. The narrative frame of much posthurricane art is an "aesthetics of disaster," which here refers to ugliness, to representations of a natural disaster's effects—its chaos, destruction, and decadence—and to effects of the societal disasters of colonization and capitalism.[5] The word *disaster* has its roots in *disastro*, an Italian word derived from the Latin *dis*, a negative meaning "bad," and *astro*, meaning "star"; *disaster* thus directly translates as "bad star" or "ill-starred," meaning an inauspicious arrangement of the stars or constellations, which bring bad luck and unforeseen misfortunes.[6] This understanding derives from sixteenth-century European cosmology, in which the universe is governed by a divine balance that the cosmos and kingdoms (or human governments) all formed part of. In other words, *disastro* meant a disordered or mismanaged occurrence. Therefore, the meaning of *disaster* is not purely a matter of nature; an aesthetic of disaster can contain meanings of nature as well as of order or governance and, thus, of human influence. This is evident in posthurricane contemporary Puerto Rican art, which conceptually suggests that people play a role in the chaos of disaster.

In using the aesthetics of disaster, recent Puerto Rican art, particularly visual art, tells the complex story of the entanglements among Hurricane María, capitalism, and colonization. What follows is a brief presentation of the posthurricane surge of art on the island and in the United States. To exemplify how an aesthetics of disaster is engaging in social catharsis, three posthurricane art pieces are discussed, specifically Gabriella Torres Ferrer's *Valora tu mentira americana* (2018), presented at "PM" at Embajada gallery, in Puerto Rico; *Tenemos sed*

[We are thirsty]: *Where did our presupuesto nacional* [national budget] *go?* (2017) by Rafael Vargas Bernard, presented at the *Focus on Puerto Rico 777* International Mall, Miami; and the *Rebuild Comerío* installation (2018), exhibited at "Defend Puerto Rico" at the Caribbean Cultural Center and African Diaspora Institute in New York by Defend Puerto Rico project-collective. This essay is far from representative of the posthurricane aesthetic expressions that are (still) unfolding before our eyes. In the pages that follow, however, I will attempt to understand how Puerto Rico's posthurricane art is thinking through the effects of María and its links with colonization, via the frame of an aesthetics of disaster. Its aim is to examine how contemporary Puerto Rican art, through social catharsis, is helping encode the story of the effects of the hurricane in its fullest complexity.

Diasporamus, by Patrick McGrath Muñíz, 2018. Oil on canvas.

The Puerto Rican arts space has been flooded with works and exhibitions that produce a catharsis that is societal, expressing not only (if at all) an individual's processing of psychopathological symptoms but also social-political discontent. The artist, the viewer, the networks of people who inform the artists and are informed by the artists and viewers all express and disseminate a message that is manifested, in this case, in the form of the contemporary art piece. In other words, art is inherently social because it is produced in a web of human connection, and it must, therefore, be culturally relevant for a given group of people. Dozens of exhibitions and hundreds of contemporary Puerto Rican art pieces using an aesthetic of disaster to portray posthurricane themes—such as *Diasporamus* (2018) by Patrick McGrath Muñíz shown above—have been displayed in Puerto Rico and in cities that have a large diasporic population (mainly in the US). To this we can add the dozens of urban art murals posthurricane in Puerto Rican and US cities. A few examples of exhibitions are "PM" (San Juan), "Defend Puerto Rico" (New York), "Puerto Rico: Defying Darkness" (Albuquerque), "Puerto Ricans Underwater" (New York), "Focus on Puerto Rico" (Miami), to name a few. Artists such as Patrick McGrath Muñíz, Elsa María Meléndez, Frances Gallardo, Gabriella Torres Ferrer, Juan Sánchez, Richard Santiago, Antonio Martorell, Adrian Viajero Román, Lionel Cruet, and many others have also responded to the hurricane with their art, capitalizing on the versatility of the narrative frame of disaster.

This outpouring of posthurricane Puerto Rican art makes evident the overwhelming drive to make sense of the naturally accelerated disaster of coloniality and the Puerto Rican sociopolitical situation. For instance, *Diasporamus* (2018) by Patrick McGrath Muñíz is an example of a piece that invites the viewer to contemplate an aesthetic of disaster, weaving together the themes of posthurricane disaster, colonization (by referring to Spanish Renaissance art), climate change, capitalism and forced displacement. When contemplating McGrath Muñíz's work, the eye immediately jumps to the center of the painting, where there is a shirtless anguish-stricken man holding a paper towel—a direct reference to the controversial incident involving US president Donald Trump when he visited the island—over a visibly saddened woman in an almost fetal position, presumably protecting herself from the peril suggested in the painting as a whole. This allegorical work functions as a conceptualist loop of gyrating local and global narratives of disaster that yarns themes of capitalism and colonization. The work brings together multiple references to Renaissance art, the depiction of vicious floods, the widely seen image of Puerto Ricans trying to get a cell phone signal after the hurricane, an explicit commentary on migration and forced displacement, iconic consumerist symbols of capitalism (Starbucks, Shell, and Yamaha, rephrased as Yamejo-di) and iconic Puerto Rican representations that range from the mundane to the satirical—for instance, the pig that travels in the small vessel, with the *jíbaro* (rural dweller), and the manatee swimming in the floodwaters, among other (sometimes hidden) narratives and symbols. *Diasporamus* capitalizes twice on the (European) Renaissance artistic forms from the past and the previous Puerto Rican histories of migration to the United States suggesting a simultaneous commentary on the two histories of colonization—those of Spain and the United States McGrath uses an aesthetics of disaster in this piece to tell the complex story of how posthurricane effects can be tied to colonization, displacement, capitalism, and Puerto Rican culture.

Now let us focus on three unexamined illustrative pieces that show how colonialism and human and natural disaster coalesce into an aesthetic assemblage. "PM" (2017), at the Embajada gallery in San Juan, Puerto Rico, was one of the first exhibitions that explicitly addressed posthurricane Puerto Rico.[7] The curation by Christopher Rivera clearly juxtaposes the gallery's chaotic "outside" with its clean "inside," which showcases the imagination and memories of those who experienced and were affected, in one way or another, by the disaster. Most of the exhibition's pieces were crafted after the hurricane, capitalizing on an explicit curation that highlighted precarious posthurricane everyday materials, yet conserving a minimalistic feel. The exhibition also produced an immersive experience into the simultaneously chaotic and calm posthurricane experience, given that the gallery was a controlled space in a Puerto Rico still lacking electricity and basic goods.

Some pieces in "PM" showcased more than others the cathartic character of an aesthetics of disaster. An example is Gabriella Torres's *Valora tu mentira americana* (Value your American lie, 2018), which displaces the room with an unembellished piece of the Puerto Rican Power Authority's failed infrastructure—a utility pole—that should be "outside" and denotes one of the main infrastructure problems in posthurricane Puerto Rico, the lack of electricity (figure 2). At the same time, Torres's piece, with its ample illumination, offers a sort of posthurricane relief to the viewer who still has no electricity. Destruction in the comfort of the gallery room, next to an air-conditioner's vent, stimulates the viewer to make sense of the destruction outside in a (relatively) calm way. The surreal experience of destruction is contained in the room, conserving the uneasiness of displacement and the danger of torn hanging electrical cables, and it allows (almost calls) the viewer to closely examine the story that this object is telling.

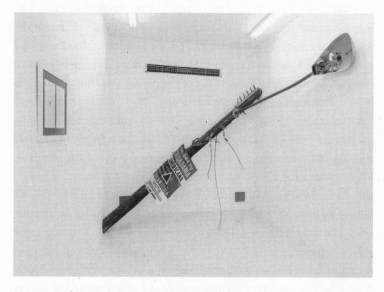

Figure 2. *Valora tu mentira americana*, by Gabriella Torres Ferrer, 2018. Mixed-media installation. Courtesy of the artist.

Utility poles in Puerto Rico are often used for political propaganda, and this is reflected in Torres Ferrer's piece, which features a flyer attached to the middle of her utility pole. The flyer reads "Valora tu ciudadanía americana" (Value your American citizenship), and it continues below with "Garatízala, vota estadidad, 11 de junio" (Guarantee it, vote for statehood on June 11), which alludes to a highly questionable plebiscite on Puerto Rico's relationship with the United States that occurred two months before the hurricane. Given how the piece contains disaster, the message in the political propaganda displays the contradiction between the perceived protection of United States citizenship in Puerto Rico and the posthurricane destruction and ongoing infrastructural disaster—focusing on electrical infrastructure—that the artist links with the futility of Puerto Rican politics, which she extends to the political (colonial) relationship with the United States. In other words, Torres's piece invades the gallery room with the gigantic collapse of Puerto Rican infrastructure, and the artist identifies the heart of the human disaster: Puerto Rican politics and its asymmetrical power relationship with the United States.

The June 11 plebiscite was highly controversial because the options available to voters did not include an "evolved" version of the current commonwealth status, the option supported by one of the major political parties. Moreover, the plebiscite suffered from massive voter abstention.[8] The ironic twist of the piece's title and its materials gesture toward a close reexamination of Puerto Rico's political and colonial status, which was exposed—to an extent—by the power of a hurricane. The irony of the piece's title, *Value Your American Lie*, also provokes the viewer to remember the political event of the plebiscite in the context of a posthurricane disaster that showcased, through the shattered utility pole, the futility of Puerto Rican politics in the face of its infrastructure's hyperfragility. In the face of bare destruction, Torres Ferrer's piece leads the viewer to question the apparent safety net of Puerto Rican political status, its US citizenship, and therefore its (presumed) preferential relationship with the federal government—causing the rude awakening that Puerto Ricans are colonial subjects—while also inviting the viewer to question the lie or ask, How did we arrive at this "lie"?

In Miami, "Focus on Puerto Rico" was an exhibition established before Hurricane María hit the island; half of its artists were stranded on the island and could not arrive at the 777 International Mall to exhibit their art. Because of this, "Focus on Puerto Rico" became more relevant as the artists in Miami began to communicate from afar their fears and anxieties about the situation in Puerto Rico. At the same time, the exhibition became a voice for Puerto Ricans in Florida, where there are almost one million Puerto Ricans, to react to and make sense of the disaster. The artists of "Focus on Puerto Rico" showed pieces that urgently engaged in a form of catharsis from another location, that of Puerto Ricans in the diaspora.

Rafael Vargas Bernard addressed a pressing issue in the hurricane's aftermath, potable water. In *Tenemos sed [We are thirsty]: Where did our presupuesto nacional* [national budget] *go?* (2017), Vargas Bernard uses the issue of potable water, corrosion, and a suggestion of a decaying water infrastructure to encourage the viewer to reexamine Puerto Rican politics, corruption, decadence, and injustice (figure 3). As in Torres's piece, disaster becomes the form in which the viewer's senses are affected (Vargas Bernard's piece causes thirst and disgust); the viewer is then led to make sense of the piece's elements as they relate to the political causes of this decadent vision. Vargas Bernard does not use the hurricane's immediate consequences—as Torres does with her utility pole, snapped by the hurricane winds—but instead calls attention to the explicit corrosion of disuse, presumably because there is no consistent running water. The aesthetic form of disaster then becomes intentionally cathartic, redolent of a decadent political infrastructure whose assemblage is a product of hundreds of years of colonization.

Tenemos sed, ¿Where did our presupuesto nacional go?, by Rafael Vargas Bernard, 2017. Mixed-media installation. Miami. Photo by On the Real Film.

Furthermore, in *Tenemos sed*, running water filters through the walls and the uneconomical water pipes, making a case for water as a metaphor for the Puerto Rican budget—the island's public funds filter out from the walls, drying up and corroding the whole system of distribution. The interplay between Vargas Bernard's piece and its conceptualization expressed in its title is prefaced by ideological political concern mobilized by the primacy of water through the affective pull of "thirst" and disgust, drawing attention to the real effects that corroded political situations have on people's lives. Corrosion functions as a near homophonic and synonymous relationship with corruption (in function) that serves as a hyperreal allegory to posthurricane experiences. The materiality of the corroded sinks, walls, and floors expresses colonial politics in Puerto Rico, particularly in a posthurricane situation, and it can be more real and more illustrative than the denotation of corruption. The cathartic aesthetic expression through "disaster" in Vargas Bernard's piece is trying to process the materiality of the individual and the collective—suggested in the fact that there are two sinks—and to urge the viewer to reflect politically on the visceral, urgent moment that Puerto Rico has lived through in the aftermath of Hurricane María.

In New York at the Caribbean Cultural Center and African Diaspora Institute (CCCADI) from February 2018, the "Defend Puerto Rico" exhibit was displayed, focusing on the posthurricane community's efforts from Puerto Rico and abroad to rebuild Puerto Rico, using an aesthetic of disaster form that highlights recovery efforts. The Defend Puerto Rico project has existed since 2016 to address myriad issues concerning Puerto Rican identity and its political relationship with the United States, particularly in the aftermath of the creation of the bankruptcy board through the .PROMESA law.[9] This transmedia art project aimed to document the complexities of Puerto Rican culture on the island and in the United States, but in the aftermath of Hurricane María the necessity to address the situation became more urgent. The exhibit, however, displayed not only the hurricane's aftermath but also the community's efforts to rebuild Puerto Rico through many grassroots initiatives. The project uses an array of media, from photography to 360-degree augmented reality captured from drones, which provides an immersive experience of what Puerto Ricans are doing to recover from the hurricane.

Rebuild Comerío: Imagine a Puerto Rico Recovery Designed by Its Communities (2018), by the Defend Puerto Rico project-collective (figure 4), is an immersive installation assembled to highlight Puerto Rico's recovery by placing Puerto Ricans from the town of Comerío at the center of the design for its rebuilding through community action via organizations such as La Maraña and Coco de Oro. The juxtaposition of posthurricane disaster and the work of recovery in Comerío displayed in photographs and drone maps, all in the mosaic shape of the island, functions as a cathartic display eliciting hope and suggesting autonomous community healing. The island's mosaic shape from representations of Comerío comments on the recursive character of individuality and community, from person to town, from town to nation, and from national again to local (Puerto Rico to Comerío). The interactive character of this installation documents the community support that people provide in rebuilding a town and the participation of an artist and many other actors in the installation. Lastly, it pieces together artist and subject or object. Its transmedia, immersive character and the indistinction between artist, authors, and subjects/objects functions as a cathartic invitation to action while mobilizing the rebuilding of Comerío and Puerto Rico as something that is possible.

Rebuild Comerío: Imagine a Puerto Rico Recovery Designed by Its Communities, Defend Puerto Rico project-collective, 2018. Mixed media.

Rebuild Comerío is above all a story that tries to capture the complex dimensions of disaster, from the narrative of chaos to inspiring stories. The collective chooses to highlight these inspiring stories of community rebuilding while revealing intentional absences of government, unless they are represented as part of the chaotic narrative of disaster, as in the destroyed power infrastructure and fragments of blue FEMA (Federal Emergency Management Agency) tarps. The highlighted story here gestures toward a long history of local adversities. Despite the crudeness of the aftermath of these natural and human disasters, there is a visceral invitation to reinvigorate community through grassroots action, highlight the Puerto Rican cultural identity as a source of resilience, and to combat cultural colonization (see, for instance, the *cuatro* guitar at top left and music records at right, among other iconic cultural symbols).

ART AS SOCIAL CATHARSIS

Social catharsis through cultural aesthetic expression channels apparently contradictory perceptions to be processed through the affective and creative character of art. Yet catharsis needs a form or a frame to be able to paint the complex picture of an apparently contradictory reality. Hurricane María laid bare a material reality that Puerto Rican contemporary art is trying to make sense of through the productive frame of disaster. The complex history of the colonial, cultural, and sociopolitical situation of Puerto Ricans cannot be fully described nor engaged in a one-dimensional manner. It is not surprising that the form or aesthetics of disaster function as a productive frame to comment on both the posthurricane dimension and the colonial dimension that permeate many spheres of everyday life, from the personal to the collective, from the cultural to the political, and from the real to the imaginary.

Two weeks after the hurricane hit Puerto Rico, I spoke with an artist about the effects of the hurricane and about the many Puerto Ricans who were moving to the United States, including many young artists. The artist said, "If I could not do art, I would leave [Puerto Rico] [*Si no pudiera hacer arte, me iba*]." I do not believe this statement speaks only of the psychological healing capacity of aesthetic discourse. It also expresses the sociopolitical capacity of a grounded aesthetics to provide the means to defy and resist colonization, capitalism, and all other dangers facing the vulnerable island.

1. An expanded version of this essay was published as: Carlos Rivera-Santana, "Aesthetics of Disaster as Decolonial Aesthetics: Making Sense of the Effects of Huricane María through Puerto Rican Contemporary Art," *Cultural Studies* (2019) https://doi.org/10.1080/09502386.2019.1607519.

2. Friedrich Nietzsche, *The Birth of Tragedy and Other Writings* (Cambridge: Cambridge University Press, 1999), 37.

3. See press release on the Museo de las Americas website, https://www .museolasamericas.org/catarsis.html.

4. Lev Vygotsky, "The Psychology of Art," *Journal of Aesthetics and Art Criticism* 30, no. 4 (1972): 570.

5. Umberto Eco, *On Ugliness* (London: MacLehose, 2007). I use ugliness in contrast to how visual arts typically capitalize on the beauty of natural disasters, as in Jenifer Presto's "aesthetics of catastrophe." Her concept focuses on an aesthetics of beauty that contains sublime visions of destruction and highlights the decadence of the beautiful with a transfigured splendor.

6. Art Carden, "Shock and Awe: Institutional Change, Neoliberalism, and Disaster Capitalism," *SSRN*, November 18, 2008, last revised May 20, 2009, https://papers.ssrn.com/sol3/papers.cfm?abstract_id=1302446.

7. The exhibition's materials included a list of almost one hundred phrases— "post-María, post-mortem, particular matter, post-melancolía"—using the acronym PM.

8. Frances Robles, "23% of Puerto Ricans Vote in Referendum, 97% of Them for Statehood," *New York Times*, June 11, 2017, https://www.nytimes.com /2017/06/11/us/puerto-ricans-vote-on-the-question-of-statehood.html.

9. The Puerto Rico Oversight, Management, and Economic Stability Act (PROMESA) is a restructuring law legislated by the US Congress in 2016 that created a financial oversight and management board to oversee any legislation that has budget implications created by the Puerto Rican government. The board is constituted by federal government appointees and therefore unelected by Puerto Ricans. For a scholarly cultural studies analysis of PROMESA, see Pedro Cabán, "PROMESA, Puerto Rico and the American Empire," *Latino Studies* 16, no. 2 (2018): 161–84.

ART AND A THRESHOLD CALLED DIGNITY

TIAGO (Richard Santiago)

> *Here lies Juan*
> *Here lies Miguel*
> *Here lies Milagros*
> *Here lies Olga*
> *Here lies Manuel*
> *who died yesterday today*
> *and will die again tomorrow*
> *Always broke*
> *Always owing*
> *Never knowing*
> *that they are beautiful people*
> *Never knowing*
> *the geography of their complexion*

—Pedro Pietri, *Puerto Rican Obituary*

The parents, just like every other adult Puerto Rican, are still holding on to the vague idea that the hurricane will go either north or south and not harm the island. I mean, a few days before, Hurricane Irma had almost the exact same trajectory, and it went up north, affecting only some regions. But this time, maybe eight hours before it makes contact, the father knows they will not escape the destruction. The family has tried to secure their Puerta de Tierra apartment to the best of their abilities. They take the dog out for one last time before the storm. Outside, the winds already feel like a category 1 hurricane, but they still have seven hours until it arrives with all its strength. Mother and father take the two boys, a six and a seven years old, into the smallest room. This is the only room where they feel a minimum sense of security. That's where they will spend most of the next twelve hours without electricity or communication . . . but at least together. Outside that room, the beast roars. The ceiling falls down and the rest of the apartment is inundated. They lose almost everything. But they are alive. And with that comes the doubting, the decision making, the survival, the ultimate intent to protect the children from any harm and confronting the storm after the storm.

That's how my journey and my family's journey began in September 2017. A roller coaster ride named displacement whose unpredictable motion has kept us vulnerable throughout.

We spent the following weeks in that little room. There, we began the early mornings worrying about the bloody bedsheets from the mosquito bites on our children's bodies, and there planned what the task at hand would be each passing day. In that little room, I cried alone after leaving all of them at the airport a month later, and there I crashed seeking rest each night after collecting what could be saved from my inundated art studio.

Our displacement took us to Chicago, where we were finally reunited on December 25. In Chicago, the Puerto Rican diaspora and their solidarity were a key ingredient to helping us cope and mitigate the traumas that María ingrained in our family. Becoming the cochair of the Arts and Culture Committee of the Puerto Rican Agenda of Chicago was an important element as well. I fully immersed myself in hurricane relief efforts focused on artists. Then the aim became to help La Escuela de Artes Plásticas y Diseño de Puerto Rico (School of Visual Arts and Design of Puerto Rico), where I used to work as a professor. This institution is the only autonomous art university in our country and has for decades been a melting-pot for the top artists to come out of Puerto Rico. Its importance in our country for the past fifty years is equivalent to an artery in our body, yet, along with the government's economic blows, the hurricane almost killed this university.

Initially, the relief efforts led me to create an event called "Rican Renaissance" on Division Street, the Puerto Rican hub of Chicago. I collaborated with multiple programs founded by the Puerto Rican Cultural Center, as well as with the help of young local artists sympathetic to our cause. Poetry, art, music, and Puerto Rican food were uniting elements in two days of coming together for the relief of the arts in our nation. The success of the event led me to return to the island, for the first time after María, to put the raised money directly in the artists' hands and contribute, as well, to the university.

An organization called the Puerto Rican Agenda of Chicago launched an intense campaign called the 3Rs for Puerto Rico, focused on rescue, relief, and rebuilding. Through multiple fund-raisers that engaged Chicagoans in our efforts and by participating in countless interviews during the following months, we helped keep the name of Puerto Rico on the table during the early stages of relief even while we were being portrayed as naggers by the US government's white elites. I barely had time to think about creating art. After María, I couldn't even conceive the idea of putting a brushstroke anywhere. It was about rescue, relief, and rebuilding. It was about my two kids and trying to find a little bit of peace for them.

But there was a nuisance in my mind. There was an annoyance that grew daily in my heart. Out of that burden I wrote the following proposal:

THE FRAILTY OF STRENGTH AND VICE-VERSA

A traveling exhibition and call to action . . .
The final moments of life for 911 people who died on, or soon after, September 20th, 2017, in Puerto Rico and whose remains were cremated by their government without official examinations, remain a mystery as do the causes of their deaths. What is clear is the government's intent to hide the real number of casualties derived from Hurricane María, the strongest natural disaster of this kind to

hit Puerto Rico in its history.

The obscurity of their deaths in this US colony is compounded by another indignity: The identities of these nine hundred and eleven people remain unknown.

THE FRAILTY OF STRENGTH AND VICE-VERSA is a last-ditch effort to bring closure and help spring forth funding for the Center for Investigative Journalism in their (and our) quest to find the truth by confronting the Government of Puerto Rico on this regard.

With this exhibit I intend to erase the anonymity of the 911 victims of Hurricane María, who were obscurely cremated by the government of Puerto Rico, by manifesting their presence through a conceptually symbolic artistic form. This show hopes to be an incisive and efficient gesture of resistance to the neoliberal forces that threaten Puerto Rico and the future of its people.

I will be creating 911 ten-inch by eight-inch Monotypes representing each person.

A Monotype is a unique image printed from a polished plate, made of glass or metal, which has been painted with a design in ink. The image is transferred from the plate onto a sheet of paper by pressing the two together, usually using a printing press.

Monotypes can also be created by inking an entire surface and then, using brushes or rags, removing ink to create areas of light from a solid area of opaque color.

The uniqueness of each Monotype intends to reference the human fingerprint and the fact that its impression affords an infallible means of personal identification, because the ridge arrangement on every finger of every human being is unique and does not alter with growth or age.

Each Monotype design will be printed on paper with a xerograph paper towel design alluding to the White House's disconnect to the loss and misery Puerto Ricans are going through. It's my way to turn one man's cynical disposition into something righteous.

And thus, another journey began. "The Frailty of Strength and Vice-Versa" soon took a life of its own. At first, friends and family members supported the project and became the first collaborators. But then, after the first presentation at the Boathouse Gallery, an alternative art space in Humboldt Park, the installation started to connect with people all on its own. Strangers were reaching out to be a part of the process. There were articles and interviews being done about it. I was invited to bring the installation in different formats to the Museum of Contemporary Art of Chicago, Connecticut College, the University of Massachusetts in Boston, and other spaces and communities. Also, it became the reason why I was invited to participate in "Aftershocks of Disaster: Puerto Rico a Year after María," at Rutgers University on September 28, 2018.

Hurricane María unveiled a lot of the real Puerto Rico to the world and to ourselves. This is also true in the arts. One night as I worked on the installation, I watched CBS news correspondent David Begnaud interview a group of Puerto Rican artists (some of whom I love very dearly) as they talked about how "María was the best thing that ever happened to Puerto Rico." They said people on the island were being sedentary, thus perpetuating the agelong misconception of Puerto Ricans as lazy, which clearly contrasts with the stories of resiliency after the hurricane. It was nauseating to hear the worst kind of neoliberal ideas being splattered on international television and these artists, knowingly or not, expressing themselves as the worst kind of disaster capitalists.

It's my opinion that the concept of tribalism is one that is connected directly to the colonial mind-set. This condition is a common one in Puerto Rico and also within our artistic community. Artistic trends divide and fluctuate through our art scenes in groups that resemble secretive Masonic lodges. It was through those groups that the information regarding art relief funds, grants, and residencies was transmitted after the hurricane; artists outside these networks lacked opportunities for aid. Artists within the nexus recommended each other and, in some cases, accepted multiple grants and fully funded artist residencies knowing there were other artists in worse conditions than they who simply were not part of their circle. María thus unmasked a lack of political consciousness and social empathy among my peers.

I feel very disconnected from the tribalist form of thinking, so in many ways, "The Frailty of Strength and Vice-Versa" became my shelter. This installation became a therapeutic safe haven for me from the moment of its conception. At the same time, it has served as a connective element among people who are still mourning our dead and as a way to ease their pain throughout that process. Its essence lies in the facts that it cannot be done without the collaboration of caring individuals and its conclusion can be achieved only with their participation. Another important element is that it is conceived to raise funds for a cause outside of myself.

Photographer: Sofi Lalonde, OLYMPUS DIGITAL CAMERA, May 2018
Boathouse Gallery, Humboldt Park, Chicago, IL

Pablo Picasso's *Guernica*, Théodore Géricault's *Raft of the Medusa*, and Vincent van Gogh's *The Potato Eaters* are but three examples of a myriad of artworks in history that depict affliction and suffering. Unlike other professions in which one's job is to report or study those kinds of circumstances, these artists were drawn to interpret them, to express the human condition of the times, and to show how those conditions affect them. Their interest in this subject matter was not economically but rather emotionally driven. Heartbroken doesn't begin to describe how I felt, and still feel, after the hurricane. That devastation fueled the artwork I've been doing since. I find it troubling and distasteful to see peers making images of other people's misery throughout their process of struggle and survival for personal gain, and I suggest to them to reach out to community leaders, nonprofit organizations, or specific individuals who are still hard at work in the recovery efforts. I urge those artists to give part of the proceeds from the sales to the people depicted in the same artwork they were "inspired" by. But mostly, I urge those artists to actively engage with those communities. It will change their lives, and their work, for the better.

In a nutshell, most people take comfort in being able to walk in the dark from their bed to the kitchen and grab any snack they choose in the middle of the night. It means there is a connection between them and the environment they have created. Home is a place of comfort and of control. That sense of order, whatever form it takes, shelters us from the chaos and unpredictability of the outside world. From fear itself.

Having lost my home and finding myself a displaced refugee, I have been decontextualized by the world. The only way now is to reconstruct. The path is a continuous search for truth. My artwork will be forever changed. Within me resonates the memory of early sunset in San Juan—before my eyes is the vivid, saturated scale of the oceanfront light and atmosphere, which breezed and thundered deeply in the shadows.

PICKING UP THE PIECES

Adrian Roman

Since October 2, 2017, a week after the hurricane made landfall, I traveled every month to Puerto Rico bringing relief aid and assisting in the recovery and rebuilding efforts. My work centered around the municipality of San Sebastian, where my paternal family is from. In the weeks after the storm, this municipality experienced great difficulty receiving assistance due to blocked roads from fallen trees, flooding, and landslides. While delivering aid to these areas, I met many people with heartbreaking survival stories that I tried to incorporate into my work.

The first community I delivered supplies to in San Sebastian was in El Culebrinas where I met an elderly women by the name of Digna Quiles. White Digna was very welcoming and invited me into her home, she was clearly traumarized from the effects of María. Digna was approximately sixty years old at the time and about five feet tall. Her husband passed away eight months prior to the storm, and she was left to live alone with her dog. She lived in a small neighborhood known as Villa Sofia, which is adjacent to Rio Culebrinas. During the storm, the river, as well as the smaller tributaries and waterways that branch off of it, swelled and rose into the streets flooding the neighborhood, including Digna's home. As the water level rose to over eight feet inside, the force of the flood water combined with constant rain and hurricane winds forced Digna and her dog our of their home. Digna was rescued by a neighbor, but unfortunately, her dog could not be rescued. When I enter Digna's home, she had no

198

furniture, visible water damage was everywhere, and she was sleeping on a child's mattress on the floor. Like most people in the wakes of hurricanes Irma and María, she had no access to clean drinking water, and was forced to collect rainwater in five-gallon buckets. My time spent with Digna inspired me not only to create the latest portrait installation *Sobrevivientes* for my Caja de Memoria Viva series, but also to create a new collection of artworks to reimagine the scope of María's damage called *Picking up the Pieces*.

Picking Up the Pieces provides an intimate view of life post-María, and the unfortunate "new normal" it has created for Puerto Rican people. The collection consists of drawings, installations, and miniature sculptures that reflect intimate moments experienced during my travels throughout the island bringing relief aid, supporting and helping with recovery and rebuilding efforts for the past year.

The installation includes a special collection of found objects I've named *PR-tifacts*, that have been donated by residents, as well as gathered from giant piles of rubbish scattered around barrios, beaches, and on properties where a home once stood. These *PR-tifacts* are more than just damaged property, they are objects that transmit the energy of the people who once owned them. They represent the destruction of the fabric of ordinary life; the memories of a life once lived, the imagination of our children, our love, our history, our pride, our faith, and all that defines who we are as Puerto Ricans.

These items demonstrate, in part, what the landscape of the island looked like post-María. The combination of artworks and found objects all woven together create portraits of individual lives profoundly touched by trauma and tragedy, travail and resilience. For weeks and months following María, families were cleaning up damage from their homes, and for many the work still continues.

Caja De Memoria Viva III: *Sobrevivientes*: Digna Quiles (detail) Charcoal on wood (outside) 2018

Caja De Memoria Viva III: *Sobrevivientes*: Digna Quiles (detail) Charcoal on wood (outside) 2018

Caja De Memoria Viva III: *Sobrevivientes*: Digna Quiles (detail) Mixed Media and PR-tifacts (inside) 2018

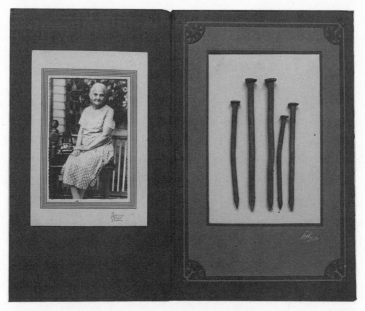

Family Portrait. Photo of Adrián's great grandmother, opposite nails from her house that was destroyed by hurricane María. Digital print, cardboard photo studio picture frame, rusted nails.

PART IV

Capitalizing on the Crisis

SINVERGÜENZA SIN NACIÓN[1]

Raquel Salas Rivera

para josé, ana, carlos y helen

vi las mejores almas de mi generación
engullidas por el colonialismo,
anestesiando sus heridas en un pozo de alcohol
con un torbellino de no sé qué totalidad pendiente.
las vi hablar de la muerte con esperanza,
llevar los cementerios de anillos,
quemar cuanta mata y matanza nos prometieron,
ocupar tierras y edificios,
odiar los ojos azules de rosselló,
escupirle en la cara a la justicia
por embustera,

estar mal y ser hermosos,

aguantar todo el dolor del mundo entre las cejas.

les toqué el pecho para que lloraran
y la ternura era un campo minado.

sin coordinación, los vi atropellar con un abrazo
el hormiguero defensivo del bienestar.

---

1   First published in *Kweli Journal*

vi que, en sus manos, la supervivencia valía un trapo,
que el linaje no cree en sí mismo si la muerte brinca citas.

vi que eran ángeles que por más de 500 años
llevan preparando el vuelo,
sin saber si queda ya cielo ni trompeta.

vi las mejores almas de mi generación perder su generosidad.
el dolor les hizo una mala jugada.

las vi colgarle el teléfono a fema
y preparar palomas mensajeras con el papeleo.
entre agotamientos, las vi construyendo techos y cerrando riñas,
enfocándose en cosas como luz, agua y entierro,
deseosos de que la tierra fuese tierra:
antígonas enterrando con pala robá.

eran volátiles como países,
dominados como países,
degollados como países.

vi que a diario desaparecían en el vuelo estático de la soledad.

fui testigo de la quema del arroz,
el giro del yagrumo.

estuvo mal lo que les pasó,
que les dieran un rompecabezas
y dijeran *toma, recoje los escombros
del dizque país.*

le explicaban a los hijos que papá se fue a un lugar
donde las calles están llenas de donas y la lotería
llega todos los meses como cheque,

pero también mataron el miedo con un range rover dorado,
formaron fila para comerse un pescado con propiedades curativas,
la montaron en barras y panteones,

hicieron lo impensable: la gran gira
por todo puerto rico llevando no la palabra de dios,
sino su carpintería,
para reconstruir un amor que aguante
lo torrencial.

también, transplantados y enormes,
eran murales sin pared. a lo alto,
mejorándolo todo con la risa,
asegurándome que la lucha
nos dará pan para el pan de cada día,
que existe cierta forma de olvidarnos que
llevamos tiempo en el bolsillo,
no en la muñeca.

de noche me soñaban alegre
en casa, en bata,
segura del mar
y de un monte que sigue engullendo
las rutas e inventos de los colonizadores.

SINVERGÜENZA WITH NO NATION

Raquel Salas Rivera

for josé, ana, carlos, and helen

i saw the best souls of my generation
swallowed by colonialism,
anesthetizing their wounds in an alcohol well
with a whirlwind of i don't know what pending totality.
saw them talk of death with hope,
wear cemeteries as rings,
burn all the plants and killings they were promised,
occupy lands and buildings,
hate rosselló's blue eyes,
spit in the face of that liar,
justice,

be wrong and beautiful,

hold (in) all the world's pain between brows.

i touched their chests so they could cry
and tenderness was a minefield.

without coordination i saw them, with an embrace,
trample the defensive anthill of well-being.

i saw that, in their hands, survival was worth a rag,
that lineage doesn't believe in itself if death skips dates.

i saw they were angels who have spent more than
500 years preparing for flight,
who don't know if there will be heaven or trumpet.

i saw the best souls of my generation lose their generosity,
played by pain.

i saw them hang up on fema
and prep messenger pigeons using paperwork,
saw them, between exhaustions, construct roofs and end beefs,
focus on things like light, water, and burial,
wanting earth to be earth,
antigones burying with a stolen shovel.

they were volatile like countries,
dominated like countries,
beheaded like countries.

i saw the daily disappearance of solitude's static flight.

i witnessed the rice burning,
the yagrumo flipping.

it was wrong what happened
when they were given a puzzle
and told *here, pick up the rubbish*
of this so-called country.

they explained to their children that their father went to a place
where the streets are full of donuts and the lottery
arrives each month like the check,

but they also killed fear with a golden range rover,

formed lines to eat fish with curative properties,
lit shit in bars and pantheons.

they did the unthinkable: the great tour
all over puerto rico spreading not god's word,
but his carpentry,
to reconstruct a love that withstands
the torrential.

also, transplanted and enormous,
they were murals without wall, up above,
making the world right with laughter,
making sure the struggle
gives us bread for the daily bread
because there is a certain way of forgetting
we keep time in the pocket,
not on the wrist.

at night they dreamt me into joy
at home, in a dressing gown,
sure of the sea
and of a mountain that keeps swallowing
the colonizer's routes and inventions.

PUERTO RICO'S UNJUST DEBT

Ed Morales

When Hurricane María hit Puerto Rico in September 2017 with category 4 destruction, the island was already reeling from austerity measures, population loss, a health care crisis, and a failing electrical infrastructure. Because of a $72 billion debt load, plus about $50 billion in pension obligations, Puerto Rico's economy had been in recession for over ten years. Schools were closing, doctors were leaving, and blackouts were an increasing occurrence. But the storm, which hit the entire island in almost unprecedented fashion, would take this precarious situation and accelerate everything, as quickly as the winds stripped its signature palm and ceiba trees.

In the summer of 2015, Puerto Rico's outlandishly high debt was declared by then governor Alejandro García Padilla as "unpayable," prompting the US Congress to devise a law, called the Puerto Rico Oversight, Management, and Economic Stability Act (PROMESA), to restructure it over the coming years. Just nine months into the existence of PROMESA's mandated Financial Oversight and Management Board (FOMB), Hurricane María's devastation exacerbated the debt crisis, as well as many of the underlying economic issues that had been afflicting the unincorporated territory for years. The FOMB was created in 2016 through PROMESA, which was passed by Congress with bipartisan support after the unwillingness of Obama and congressional Democrats—with the exception of Bernie Sanders and Robert Menéndez—to call for debt relief or debt forgiveness. It can be argued that US politicians, who receive large contributions from

Wall Street hedge funds, were disinclined to press for debt relief or forgiveness because it would be unfavorable to some of their main benefactors.

Puerto Rico's netherworld status—not a sovereign nation, not a US state—prevented it from using tactics like those available to Greece and Argentina, which were able to receive loans from the IMF, to renegotiate them, and also to adjust their currencies. Puerto Rico, in addition to having to use the US dollar as currency, does not have access to international courts of law and is further punished by a maritime restriction under the 1917 Jones Act which prevents ships from docking if they were not constructed in the United States, are not flying the US flag, and lack a certain number of US personnel; this drives up the cost of several consumer items.

Both the creation of the debt and the inadequate response to the hurricane could be seen as products of (1) the United States colonial and racist treatment of Puerto Rico and its other territorial possessions, which set them up to be exploited economically and denied the full rights of US citizens, and (2) the willful neglect of collaborators among Puerto Rican economic and governmental elites, who conspired with deregulated Wall Street banks and institutions to create financial instruments that would delay payment by borrowing more money and accrue higher interest rates in the long run. These same elites offered little resistance to the Trump administration's policy of neglect, from helping to cover up the number of dead caused by the storm, to an appalling lack of criticism of sluggish efforts from FEMA and inadequate appropriation of emergency funds, until almost a year after the storm, when it was too late to mitigate much of the physical and emotional toll.

WHAT CAUSED THE DEBT

Puerto Rico's massive debt was caused primarily by its colonial relationship with the United States, which, by refusing to consider its incorporation as a state after acquiring it as war booty after the 1898 war with Spain, used it as a laboratory for the excesses of unfettered capitalism. Following a pattern of indebting or speculating on the debt of several Caribbean island nations, including Cuba, Haiti, and the Dominican Republic, Puerto Rico was established as a place where US corporations could set up shop without being taxed. Because the island had no national sovereignty, it was restricted from making trade pacts with neighbors, unable to develop a national economy that repatriated profits, and forced to go into deep debt in the form of bond issuances that were needed just to maintain operating expenses for governmental agencies and essential services.

While borrowing actually began in earnest the 1970s in the wake of that decade's OPEC-induced recession, it exploded in the 2000s as the island commonwealth—more aptly described as an unincorporated territory—was thrown into recession because of the phasing out of tax breaks to US corporations under section 936 of the US Internal Revenue Code. The neoliberal reaction to FDR's welfare state—fostered by the Clinton presidency—had enabled the deregulation of the financial sector, unleashing massive Wall Street speculation, the creation of financial instruments that bundled risky debts to be bought and sold, and unscrupulous behavior by big banks and financial institutions, which restructured debts and charged exorbitant underwriting fees. These actions caused Puerto Rico's debt load to increase sevenfold from the year 2000 to present; it has sold $61 billion in bonds to Wall Street since 2006, the final year of the section 936 tax breaks.

After the municipal bond market took off in the late 1990s, investors in Puerto Rican government debt were a mixed bag, including mutual funds like Oppenheimer Funds and Franklin Templeton. But since Puerto Rico's bonds were devalued to junk status in 2014, a growing presence of hedge funds emerged. These speculative investments avoided regulatory oversight and were solely motivated by the opportunity for windfall profit, hoping to capitalize on distressed economies like Puerto Rico's.

As the crisis grew in scope, Franklin Templeton and Oppenheimer Funds successfully challenged García Padilla's 2015 attempt to legislate Puerto Rico's right to a debt restructuring that would resemble bankruptcy. That legal test encouraged more hedge funds and their more speculative cousins, vulture funds, to hold increasing amounts of up to 50 percent of the island's debt. This tendency was signaled in March 2014, when Barclays brokered a municipal bond issue of an unprecedented $3.5 billion by the island's Government Development Bank.

Vulture funds, hedge funds' more extreme counterparts, target debt that is distressed or in danger of default in troubled economies, hoping to cash in on settlements after buying the debt for pennies on the dollar. They work to gain leverage in the debt restructuring process by insisting on repayment at full face value. Given Puerto Rico's recent history of privatizing its airport and highway toll collection system, it was already vulnerable to further sell-offs—even its prized university system—as concessions to the vultures. As Puerto Rico's debt became more distressed, vulture funds bought up billions of dollars' worth of bonds at reduced prices, betting on recouping them through legal actions in courts that did not allow Puerto Rico full bankruptcy protection.

In 2016 two investigative efforts, one by an activist group called the Hedge Clippers and another by Puerto Rico's Center for Investigative Journalism (CIJPR), tried to methodically identify the hedge- and vulture-fund owners of the debt. Among them are BlueMountain Capital and Stone Lion Capital, which includes Paul Tudor Jones, who founded the Robin Hood Foundation, a nonprofit founded by Wall Street high-rollers that uses philanthropic solutions to fight urban poverty. Many of these hedge fund operators have also been targeted by activists because of their ties to New York governor Andrew Cuomo and their influence on pro-landlord and pro-charter-school legislation.

Another salient figure is John Paulson, who has not only bought $120 million in bonds but also invested in the island's largest bank and several major hotel properties. In 2014, along with Alberto Bacó Bagué, García Padilla's secretary of economic development and commerce, Paulson pitched Puerto Rico as a tax haven for renegade billionaires, saying it can become the "Singapore of the Caribbean."

Under the previous administration of Governor Luis Fortuño—a fiscal conservative who was in office from 2009 to 2013 and who cut government jobs and attacked unions—the Puerto Rico legislature passed two laws, the Export Services Act and the Individual Investors Act, also known as Law 20 and Law 22. The first law gave hedge fund managers a flat 4 percent tax rate as an incentive to move operations to Puerto Rico, while the second offered investors complete tax exemptions on dividends, interest, and capital gains, provided the investor lived on the island for half the year. García Padilla's more moderate administration had embraced this policy in a desperate attempt to maintain outside investor interest.

The CIJPR's list featured a chart that illustrated how many of these hedge and vulture funds also had a hunger for investments in troubled places like Greece, Argentina, and Detroit, which famously declared bankruptcy in 2013. Three funds, Aurelius Capital, Monarch Alternative Capital, and Canyon Capital, have been involved in all four, (including Puerto Rico) while Fir Tree Partners and Marathon Asset Management, among others, hold a trifecta of Puerto Rico, Greece, and Argentina. Paulson had also invested heavily in Greek banks.

The disingenuous smoke-and-mirrors tactics of many of these hedge and vulture funds should make their legal claims ethically and morally dubious, but such considerations count for little in the high-stakes face-off of debt restructuring as overseen by court proceedings mandated by Title III of the PROMESA law. For instance, Double Line Capital's Jeffrey Gundlach had more than doubled his holdings of junk-rated bonds in May 2015. In an interview with Bloomberg News, Gundlach said he compared the investment potential in Puerto Rico's debt to US mortgage markets in 2008, evoking the perfect-storm conditions that helped set off the Great Recession.[1]

PRELUDE TO LA JUNTA

By the summer of 2015, the García Padilla administration had commissioned a report from Barbara Krueger, an economist who had previously overseen the IMF's response to debt crises in Argentina and Greece. She was hired by the commonwealth's government to produce her report with two other IMF functionaries, sending mixed messages about how to be "fair" while "balancing" the concerns of wealthy investors with everyday citizens who are stuck on the wrong side of the balance sheet, doomed to an ongoing global project of exclusion and increasing inequality.

At a hearing held that summer at Citigroup headquarters in New York, Krueger launched into a frenetic PowerPoint presentation highlighting her supply-side suggestions for economic restructuring. "Puerto Rico's minimum wage, at $7.25 an hour, is 88 percent of its median wage," she said clinically (in fact, her own report, as well as the Bureau of Labor Statistics, put the figure at 77 percent). "Most economists conclude that half that amount would be beneficial." One major problem for the island, according to Krueger, is that it is competing with neighboring Caribbean islands, whose wage scales are lower than Puerto Rico's, which is set by US minimum-wage law. She also suggested that Puerto Rico's welfare payments are "very generous relative to per capita income," causing a disincentive to work for minimum wage. This ignores the fact that, as the pro-statehood party correctly argues, such payments are capped at levels significantly lower than what residents would receive if Puerto Rico were a state; the island's lack of full entitlement is one reason Puerto Ricans are second-class citizens.

These assessments lay bare the fading promise that lies within Puerto Rico's status as an unincorporated star in the US safety-net orbit. In a way, the federal entitlements provided to Puerto Ricans were part of the bargain to offset their watered-down citizenship, at least during the postwar boom years. But now that neoliberal free-trade models reigned amid a Great Recession still weighing significantly on the mainland, the United States could no longer afford to prop up the island's living standards and had to defer to its deregulated banking system's needs, which are to place the island in its "proper" Caribbean context, adjusting the wages of the average labor to that of its surrounding islands and the rest of Latin America.

All this was setting the stage for the establishment of the FOMB, which would be put into place as the latest of a series of financial control boards that had been employed in the United States since the late nineteenth century. Perhaps the most famous of these was the one imposed on New York City in the mid-1970s after President Gerald

R. Ford refused federal funds to bail the city out of its fiscal crisis, but more recently Washington, DC, Philadelphia, and Detroit had been sites for this sort of supervision.

Puerto Rico differs from these cities politically and economically, however, because the island lacks voting representation in Congress as well as voting rights in presidential elections, so it is financially compromised to begin with—having gone through ten years of recession but having none of the recuperative ability to draw investment as do those US cities. The FOMB in Puerto Rico was going to be less of a way for Puerto Rico to get back on its feet, as advertised in Washington and San Juan, and more in the way of negotiating higher returns to vulture fund investors.

Many of these funds are now organized into the Government Development Bank Ad Hoc Group (hereafter Ad Hoc Group), a coalition of vulture funds with $4.5 billion in Puerto Rico's Government Development Bank bonds, and the so-called PREPA Group, which holds the debt of PREPA, the government electrical utility company. The lists compiled by Hedge Clippers and the CIJPR were based on reports in the business press and were confirmed by accessing public records and in some cases verifying with the hedge funds themselves, but they are not definitive. At the time the CIJPR filed a lawsuit in a San Juan court against García Padilla and Government Development Bank chair Melba Acosta to force them to disclose the list of hedge funds that hold the bonds, the members of the Ad Hoc Group who has been meeting with her and other public officials, and the written conditions for renegotiating payment terms as well as future bond sales.

That summer the CIJPR published a bombshell article that disclosed that hedge- and vulture-fund representatives "visit the offices of legislators at the Capitol constantly."[2] According to the report, they had sometimes been accompanied by lobbyists like pro-statehood party members Kenneth McClintock and Roberto Prats, who also happened to be a major Democratic fund-raising bundler and at the time chair of the Democratic Party in Puerto Rico. Yet the article's sources for these revelations, commonwealth party senator Ramón Luis Nieves and Melba Acosta, for the most part, claimed to not remember or know the names of the hedge funds or their representatives.

"[Acosta] was not even concerned about vulture funds and hedge funds," said Carla Minet, one of the coauthors of the article. "She was more worried about Oppenheimer and the mutual funds." The judge hearing the lawsuit found that some of the CIJPR's requests for information were valid and instructed them to re-petition the Government Development Bank with a specific request for a list of bondholders and the amounts owed.

Several reports written about the accumulation debt, by groups like the Action Center on Race and the Economy, as well as one commissioned (though severely underfunded) by García Padilla's government, have pointed out that the accumulation of the $72 billion debt involved illegal and extraconstitutional practices.[3] From the "payday loan" structure of what are called capital appreciation bonds, which resemble the adjustable-rate mortgages that helped cause the 2008 financial crisis, to the massive fees charged by banks like Goldman Sachs to underwrite new loans, Puerto Rico was exploited because of its "foreign in a domestic sense" status. Its triple-tax exemption and its lack of ability to declare bankruptcy emboldened investors because they would likely win in court when they brought their cases.

That summer there had been a growing movement in Puerto Rico to carry out a citizens' debt audit, but the point of PROMESA seems to have been all along to avoid a serious one. A report issued by FOMB in 2018 identified many of the reasons for the accumulation of the debt but did not ask hard questions of anyone involved in the deals that created it.[4] Another factor must surely be the intertwining of Wall Street interests with the governmental elites in both Washington and San Juan. One case in point is hedge fund speculator Marc Lasry.

Lasry is perhaps more famous at the moment as the co-owner of the NBA's Milwaukee Bucks, a development that was so energizing for Wisconsin governor Scott Walker that he signed a bill in 2015 subsidizing a new arena for the team that, according to the *New York Times*, would "cost the public twice as much as originally projected."[5] But it turns out that Avenue Capital was one of the vulture funds that owned some of Puerto Rico debt and was aligned with Candlewood Investment Group, Fir Tree Partners, and Perry Corp, as part of the Ad Hoc Group.

In 2017 the Ad Hoc Group hired the law firm Davis, Polk & Wardwell to represent it in PROMESA's Title III process, hoping to recoup its investment and avoid either the government's debt-restructuring proposals or a move by Congress to change federal law to allow the commonwealth/unincorporated territory/colony to declare bankruptcy. Ironically, this is the same law firm that helped orchestrate the US government's bailout of AIG, the bad-mortgage debt-swapping machine at the center of the 2008 recession. So, the same law firm that pushed for the AIG bailout was gearing up to force Puerto Rico to pay up, while Obama, who also favored the AIG bailout, had agreed that PROMESA represented Puerto Rico's best hope.

Lasry was perhaps the kind of benefactor—someone who raised $500,000 for Obama's last campaign—whom Obama and the Democrats might have thought they should keep happy. After all, Lasry was Obama's choice for ambassador to France in 2013, but unfortunately "had to remove his name from consideration after a close friend was named in a federal indictment for playing in a poker ring with alleged ties to the Russian mafia." In May 2016, Lasry threw a $2,700-a-head fundraiser for Hillary Clinton, while assuring Bloomberg TV viewers that she is "moving a little bit to the left."[6]

Lasry's ties to big Democratic politics had gone back many years. A March 2010 feature in the *Wall Street Journal* described him lunching with then White House chief of staff Rahm Emanuel, in part to advise Emanuel on whether banks would resume lending again in the wake of the 2008 crisis.[7] According to a 2012 *New York Times* article, "About 50 people paid $40,000 each to crowd into an art-filled room" in Lasry's apartment to hear Obama and Bill Clinton speak.[8] Last decade, Lasry's Avenue Capital even famously employed Chelsea Clinton, whose husband had more recently flopped in making bad investments in Greece while heading his own hedge fund.

Lasry, who was once a humble UPS driver and whose parents convinced him to go to law school, was apparently a gambler at heart, capable of rolling the dice with anyone in the global Wall Street hedge-fund casino dice game—as well as actual casino owners, like then Republican candidate and now President Donald Trump. This partnership, which stretches back to Trump's Atlantic City casino bankruptcy in 2009, eventually resulted in Lasry buying him out and becoming the chairman of Trump Entertainment Resorts in 2011, a post Lasry eventually resigned. The stories about Lasry in the business press described him as the "don't call him that" vulture-fund investor; the optimistic gambler who "bets" on economies like those of Spain or Greece to "recover," and then profits from that. A 2012 *Bloomberg* story described a regular poker game he had with other hedge-fund managers; one colleague assessed him as "good at figuring out what the odds are. He's willing to take moderate risk."[9]

It was hard to believe that someone worth $1.87 billion, according to a 2016 *Forbes* estimate—presumably an indication of good business sense—would believe that economies that are in a "death spiral" would miraculously recover. It's more likely that rather than believing in a Puerto Rican economy that had shown no signs of growth for so long, and whose economy was largely driven by government employment, Lasry bet that its inability to declare bankruptcy would yield a higher return once it defaulted. Avenue Capital was one of many vultures that began hovering over Puerto Rico in late 2013, when its junk-leaning bonds caused credit analyst Richard Larkin to say of the vultures, "They can smell the blood and the fear."[10]

A, July 19, 2016, *Wall Street Journal* article on the Puerto Rico crisis provided a possible explanation for his interest in pushing back against bankruptcy or a debt restructuring that would give relief to Puerto Rico and its people: low default rates in corporate debt had led such distressed-debt specialists to instead focus on cash-strapped governments like Greece, Argentina, and Puerto Rico.[11] But while prices of Greek and Argentine bonds bottomed out at less than 20 cents on the dollar at the height of their debt crises, much of Puerto Rico's debt still trades between 50 and 70 cents, according to MSRB (Municipal Securities Rulemaking Board) data. That meant if the hedge funds are to turn a profit, they would need to recover more in a Puerto Rico restructuring than speculators in the Greek and Argentine defaults did.

If turning that profit entails austerity measures like lowering wages and cutting pensions and laying people off, thus pushing toward a human rights catastrophe—well, a smart investor has to get those 50 to 70 cents on the dollar somehow. It remains to be seen whether Obama really had Lasry's back in his capitulation to the PROMESA process. But one thing was clear: if investors had their way, a chunk of the money the Puerto Rican people would be paying back might be owed to an unrepentant gambler who had a privileged, back-door channel to the White House.

As the Title III process played out in Judge Laura Taylor Swain's pseudobankruptcy court in San Juan, more and more of Puerto Rico's current and potential holdings were being sold off to satisfy bondholders. Some of these bondholders are modest investors—including many Puerto Rican citizens themselves—but they would wind up losers in the settlements approved by Swain's court in February 2019. The bulk of the profit was assigned to hedge and vulture funds, the entities enabled by decades of deregulation and with the complicity of Democrats, Republicans, and their collaborators in San Juan. The COFINA settlement, which restructured $17.6 billion of the debt, used the same bond-swap or maturation-delaying capital appreciation devices as the ones used in the massive emissions of the mid-2010s.

Puerto Rico's dubious territorial status, not sovereign enough to negotiate with a global entity like the IMF, unable to access a true bankruptcy proceeding, has made it a captive for decades of austerity and severe exploitation. This is why a true accounting of the debt is necessary if only to attain the moral high ground to force reform of the colonial PROMESA process. Yet the debt crisis itself has been used to stigmatize Puerto Rico as an unworthy candidate for direct US aid, with Medicare benefits and food aid for the poor being sharply cut or threatened with elimination by the Trump administration. The portrayal of Puerto Rico as an indebted territory located in the languid tropics feeds into the same sort of stereotypes that vilified "welfare queens" and "super-predators." Trump's famous declaration on Twitter in October 2018 that Puerto Rico did not deserve further funding because of "inept leaders" trying to use recovery funds to pay off the debt not only is untrue but also displays a profound misunderstanding of how the austerity policies being imposed by the

FOMB are designed to make Puerto Rico residents pay an inflated price for their government's debt for the next forty years. In the current atmosphere of scapegoating and vilifying various marginalized groups—from African Americans to women to LGBTQ communities, on the basis of their racial and gender-based "difference"—the attacks on Puerto Rico will no doubt continue, and their "unworthiness" to receive as much recovery aid as Texas and Florida did after their storm crises will continue to be explained through depicting them as fiscally irresponsible second-class citizens, made invisible by over a century of myopic, insidious neglect.[12]

1. Mary Childs and Kelly Bit, "Gundlach Sees Puerto Rico Like Mortgages in 2008 Crisis," *Bloomberg*, May 4, 2015.

2. Joel Cintrón Arbasetti, "Funcionarios se lavan las manos tras autorizar deuda que podría ser illegal" Centro de Periodismo Investigativo, June 14, 2016, http://periodismoinvestigativo.com/2016/06/funcionarios-se-lavan-las -manos-tras-autorizar-deuda-que-podria-ser-ilegal/.

3. Carrie Sloan and Saqib Bhatti, "Wall Street's Power Grab in Puerto Rico," Action Center on Race and the Economy, May 25, 2017; Saqib Bhatti and Carrie Sloan, "Goldman's Strong Man in Puerto Rico," Action Center on Race and the Economy, April 28, 2017; Carrie Sloan and Saqib Bhatti, "Puerto Rico's Payday Loans," *ReFund America Project,* June 30, 2016.

4. The Financial Oversight and Management Board for Puerto Rico Independent Investigator "Final Investigative Report," Kobre & Kim LLP, August 20, 2018.

5. Michael Powell, "Bucks Owners Win at Wisconsin's Expense," *New York Times*, August 14, 2015.

6. Annie Karni, "Clinton Adds Fundraisers," *Politico*, May 4, 2015; "Marc Lasry: Hillary Clinton Is Moving a Little Bit to the Left," *With All Due Respect*, May 19, 2015, https://www.bloomberg.com/news/videos/2015-05-19/marc -lasry-hillary-is-moving-a-little-bit-to-the-lefthttps://www.bloomberg.com /news/videos/2015-05-19/marc-lasry-hillary-is-moving-a-little-bit-to-the-left

7. Mike Spector, "Avenue Capital's Investor in Chief," *Wall Street Journal*, March 27, 2010.

8. Jackie Calmes, "Clinton Supports Obama at New York Fundraisers," *New York Times*, June 4, 2012.

9. Gillian Wee, "Lasry Sees Europe Bankruptcy Bonanza as Bad Debts Obscure Assets," *Bloomberg*, February 14, 2012.

10. Michael Corkery and Matt Wirtz, "Hedge Funds Are Muscling Into Munis: Sharp-Elbowed Investors See Potential in the Stodgy World of Municipal Debt," *Wall Street Journal*, November 11, 2013

11. Matt Wirz and Aaron Kuriloff, "Mutual Funds Are Front and Center in Puerto Rico Talks," *Wall Street Journal*, July 18, 2015.

12. This chapter was adapted from "How Hedge and Vulture Funds Have Exploited Puerto Rico's Debt Crisis" and "Is an Obama Donor Tying the President's Hands on Puerto Rico's Debt Crisis?," which appeared on July 21 and August 19, 2015, in *Nation*.

PUERTO RICO'S DEBT IS ODIOUS

Natasha Lycia Ora Bannan

The images of Hurricane María ravaging Puerto Rico in September 2017, and the ensuing humanitarian crisis, flooded the media for over a year and provoked widespread outrage at government neglect, as well as empathy and solidarity for those struggling to recover their access to basic services. Months went by, and the electricity wasn't restored; the health care system was on the perpetual verge of collapse; public sources of water were being contaminated, affecting nearly two million Puerto Ricans; and thousands began fleeing the island en masse to access shelter, health care, and jobs. As all this took place, the news turned a critical eye to what was impeding recovery efforts. What most reporters and solidarity brigades found was a crippled infrastructure that, long before the hurricane, had been broken as a result of Puerto Rico's $73 billion debt. The island's main agencies responsible for public health, safety, and sanitary services were vastly underfunded, indebted, accused of corruption, or notorious for their inefficiency and systemic neglect. Some of them were also billions of dollars in debt, including the Puerto Rican Electric Power Authority, which couldn't afford to properly restore and maintain the electrical grid, having suffered a fatal blow from the hurricane. The economic crisis propelled the humanitarian crisis further into despair, while the humanitarian crisis deepened the economic depression that had already gripped Puerto Rico.

Indeed, there has been a crisis in Puerto Rico for years, and only recently has it surfaced to become more visible as a result of Hurricane

María and the international community's watchful eye on the fatal government response. We think of the crisis as an economic one that is crippling Puerto Ricans' ability to come out from under the island's oppressive $73 billion debt, not including its pension obligations. Or we think of it as a humanitarian crisis caused by Hurricane María and exacerbated by gross negligence and incompetence by the local and federal governments. The crisis was, however, created long before that and has always been as urgent, and that is the crisis of colonialism. The heart of the economic crisis is political. That is why Puerto Rico cannot achieve a more autonomous future and sustainable economy—even though it may have the tools to do so—without fundamentally changing its colonial relationship with the United States and, thus, loosening the economic stranglehold that the colonial debt has on the island. The first step in this process is to examine the debt's legality: How was it acquired, and was it acquired in accordance with the law, including Puerto Rican law? This includes examining the debt's morality and fairness according to the doctrine of *odious debt*.

This doctrine is based on principles of equity, not necessarily black-letter law. It recognizes that there are illegitimate debts that originate from either odious borrowing or odious lending practices. Either way, the result is a debt burden for a country that ends up having a destructive impact on people's lives, corroding human rights, and seriously diminishing social and economic possibilities for future generations. Historically, the doctrine of odious debt comes from the context of transitional justice, in which a government that oppressed and subjugated its people gives way to a new era of democratic rule, or at least an era in which new political leadership emerges, signaling an ideological shift and change in governance. The new government then repudiates the offensive practices that were exercised by its predecessor, which led to the country's indebtedness. These repressive debt practices usually involve a dictator or authoritarian regime that borrows to finance mechanisms of state repression often while regime members personally profit from the borrowing. Traditionally, what is considered odious in these debt practices is the regime itself and what it represents, but not necessarily the series of individual transactions for every loan. In the context of Puerto Rico, however, it is important to look at both the overall political "regime"—including the colonial government and the federal government, which created the conditions for a nation to borrow $73 billion—and the individual financial transactions that banks and the government negotiated and entered into. This is important because there have been multiple administrations, both local and federal, as well as multiple banks and hedge funds that have engaged in predatory lending. While the model of lending and borrowing didn't vary much, the amount varied by administration, ·resulting in unconstitutional debt service and borrowing.

The doctrine of odious debt raises the questions: What would be a just result for a people overcharged for debt that is inherently not theirs, and what is the behavior we want the law to either correct or encourage as a question of justice? To argue that a debt is odious is to argue against paying it and to limit a country's general obligation to pay its debts. For a government making the transition from an abusive, antidemocratic, and dictatorial regime to a democratic regime in a postconflict society, discharging odious debt means releasing the grip that international banks and other nations may have on it, impeding the socioeconomic progress necessary for that country to shift from having an economy dependent on corrupt practices to one that is transparent and accountable to its people.

There are various forms of illegitimate debt, of which odious debt is one. Odious debt covers the spectrum of odious borrowing and lending, including war debt, subjugated debt (discussed below), regime debt, and, in more modern times, unsustainable debt. The principles that are central to odious debt allow a government to break with the colonial or fraudulent bonds that have repressed people and to declare a debt illegal and immoral, either in part or in its totality. If a debt is claimed to be odious, investigators examine whether it was accumulated, used, or even paid in a way that financed corrupt or unlawful practices while either directly keeping a people under oppressive rule or undermining their interests and needs. A people can move to cancel the debt once an adjudicating body declares the debt unlawful or illegitimate, or the international community recognizes it as such, creating political pressure to hold debt negotiations.

Ironically, this argument was first asserted by the United States against Spain after the Spanish-American War. Having acquired the former Spanish colonies of Cuba, Puerto Rico, the Philippines, and Guam, the United States refused to assume Spain's debt in Cuba, which, it argued, had been accumulated to repress the Cuban people and to maintain Spain's colonial rule. Arguing that such a debt should not be passed on to a new regime (even a new colonial regime), the United States claimed that paying Spain for indebting Cuba would only reward brutal behavior that should be condemned. Thus, in the Treaty of Paris the United States did not assume the Spanish debts, nor did Cuba after securing full independence. Over a century later, it is now one of those same colonies that asserts the very same argument against the economic centers of power in the United States that maintain—with the conspiratorial help of the federal Fiscal Control Board—a perpetually subjugated nation that cannot find its way out from under an oppressive debt.

An odious-debt analysis looks at a few factors, the first being whether the people in whose name the debt was accumulated benefited from the money borrowed. Presumably, when a nation borrows money, it is to help fund its public budget, build infrastructure, or expand services to its citizenry, whose future generations will carry the burden of repaying the debt. But what happens when people not only do not receive the benefits of debt but are also forced to face a perpetual deterioration of human rights and public services, culminating in a new austerity regime? Put simply, it is as if a mother took out student loans in order to pay for her education. If she is a low-wage worker, she ends up paying back those loans for twenty years at a high interest rate because she is a "high-risk" debtor. She will eventually pay triple the cost of her tuition to banks that pocket all that extra money, money that she cannot use to improve her own economic standing, such as by acquiring better housing or building her savings. Perversely, she will also have less money to spend on her own children's education. Her children, as a result, essentially end up paying for their mother's education while depriving themselves of the same opportunities. While the mother is paying the banks back every cent plus interest, her government uses public funds to guarantee the bank the value of the loan to her, but it does not guarantee the mother or her children their fundamental right to a free and quality education.

When we look at whether the people of Puerto Rico bene- fited from the debt borrowed by their government over a period of decades, we have to ask if there was really an improvement in public services. For decades, the economy of Puerto Rico has been in steady decline until reaching its tipping point—an economic depression. Under the administration of Governor Luis Fortuño (2009–13), thousands of public employees were laid off, public services were cut, public offices were closed, and the rights of public workers were stripped away—even as the Puerto Rican government amassed huge amounts of debt. The debt did not improve but rather worsened the island's economic problems.

A second factor that odious-debt analysis looks at is whether the lender knew that the funds would be used for morally questionable ends. In addition to this question, however, indebted nations are increasingly demanding a consideration of the illegitimate *contract- ing* of debt, not just of how the loan money will be used. In the case of Puerto Rico, the government signed contracts that exploited the island's impoverished economic conditions, and these contracts were pushed, promoted, and entered into by hedge funds (often referred to as vulture funds for their aggressive practices of buying distressed debt and holding debtors in a type of debt bondage). These funds continued to buy debt even after Puerto Rico's bond rating had been lowered to junk status. They also began to charge exorbitant interest rates, which is precisely what institutions do when they consider the debtor high risk, meaning they might be unable to repay the loan. These enormously high interest rates and stringent terms of borrow- ing are intended to minimize the creditor's risk while forcing the debtor to meet impossible and expensive obligations, which means using its public budget funds to pay interest to Wall Street banks (think of pawnshops or payday check-cashing stores). This can get so expensive that governments, such as the one in Puerto Rico, end up borrowing more money just to pay the bank's interest rates without ever touching the principal. This type of predatory lending raises con- cerns about the illegitimate contracting practices of creditors under domestic law and equitable defenses such as unjust enrichment, in addition to an odious-debt analysis under international law.

This practice is called odious lending and has become part of odious-debt analysis, using the concept of *unsustainable debt*. Debt is considered unsustainable and should therefore be canceled if repayment (not accrual) requires that governments violate human rights or deprive their citizens of basic needs. Odious lending practices like those of many hedge funds have been repudiated by international organizations that promote socially responsible investment and lending, including charging fair and reasonable interest rates, promoting transparency in financial transactions, adjusting debt for changed circumstances that prevent the debtor from paying all or part of what is owed, complying with labor and environmental standards, and promoting a restructuring or debt payment plan that prioritizes the fulfillment of human rights. The devastation caused by Hurricane María in 2017 alone fundamentally changed the economic landscape of Puerto Rico and its already-limited capacity to pay its debt. After being hit by the worst hurricane in over a century, resulting in nearly five thousand deaths, hundreds of thousands of destroyed homes and property, and a total blow to the nation's weakened energy grid, the priority of the island's government must be to provide basic services to its people, who over a year later continue to suffer the long-lasting effects of the hurricane. Insisting on debt repayment when the government cannot afford to update its electrical grid or establish renewable-energy systems, or to fund public schools, higher education, and medical facilities, is precisely the type of odious lending practice that must be contested.

Although former governor Alejandro García Padilla declared the debt unpayable in 2015, there was never a call to cancel it. Instead, Puerto Rico passed a domestic law that would have allowed the island to restructure its debt. But the day the governor signed it, several hedge funds sued to have it declared unconstitutional.[1] They have been clear all along that they are uninterested in negotiating to reduce the debt repayments they have always sought, by any means necessary, including asking a court to find that the Puerto Rican government must repay not just the principal but also the exorbitant interest before it can fund its own public budget. Puerto Rico asked the federal government for funds to help with its payments, as with the Wall Street banks that received federal money after the foreclosure and housing crisis of 2007–8. House Speaker Paul Ryan, however, publicly said that Puerto Rico was requesting a "bailout" and emphasized that no "American tax dollars" would be spent to help Puerto Rico. Instead, Congress passed the law known as PROMESA as a debt-restructuring mechanism, but that came at the cost of a federal Fiscal Control Board.

The concept of unsustainable debt allows us to recognize that debt can be odious in many ways and that demanding debt repayment can also break laws, including human rights laws. When we look at what laws the government is obligated to uphold, we are not only talking about contracts and private rights. In fact, a government's primary purpose is to respect, protect, and fulfill the human rights of its citizens. Not even in times of war can human rights be suspended, much less in moments of capitalist crisis. The US government, as well as the government of Puerto Rico as its colonial territory, have the duty to ensure the ongoing realization of these most basic rights. In the hierarchy of rights and law, the vulture funds and Fiscal Control Board are mistaken; it is not the debt contracts that have legal priority but rather the people's human rights, which have been guaranteed in international law and in several instances ratified domestically. That means that any federal funds that were allocated to Puerto Rico for disaster relief after Hurricane María cannot enter the general stream of revenue for the island's budget, from which creditors will try to access them for debt service. The United Nations independent expert on foreign debt and human rights, Juan Pablo Bohoslavsky, has said as much: "Ensuring financial stability, controlling public debt and reducing budget deficits are important goals, but [they] should not be achieved at the expense of human rights. The population cannot be held hostage to past irresponsible borrowing and lending. The economy should serve the people, not vice versa."[2]

Demanding debt repayment also requires an in-depth understanding of how the debt was accumulated and whether this was done lawfully. Achieving such an understanding is perhaps the greatest barrier to the government of Puerto Rico in arguing that the debt is odious and that it can therefore refuse to pay it. In addition to the deep fear that the island will never regain access to the financial markets—which the government has relied so heavily on, leading to the island's economic depression—there is little to no political will to investigate and audit the debt incurred by current and former administrations. Yet for years people in Puerto Rico have been demanding a comprehensive audit of the debt, which would include assessing its legality and identifying all responsible parties (both public and private). The Frente Ciudadano por la Auditoria de la Deuda (Citizens' Front to Audit the Debt), a broad-based alliance of sectors across society, is leading the demand for a forensic debt audit that would help Puerto Ricans understand what they are being asked to pay, how the debt was used, who is responsible for accumulating the debt, and what *should* be paid. Since it is the citizens of Puerto Rico who are being left with the debt bill, even though they are not the ones who ran up the charges, they are seeking to clarify what is theirs, if anything. A debt audit can be just the first step in declaring debt odious and can play an integral role in identifying any financially risky or unethical transactions.

Puerto Rico's debt must be declared odious, because there is nothing more odious than being a colony. A fundamental principle of the odious debt doctrine is that the political context in which debt arises is a crucial factor in determining its legitimacy. This was key to the United States' repudiation of Spain's colonial debt. Debt that is accumulated to further colonization or to maintain colonial subjects is referred to as *subjugated debt*. When we consider why Puerto Rico was forced to borrow so much money with disastrous terms and to turn to nefarious creditors and companies with long histories of causing economic disasters, it's because as a colony Puerto Rico has been locked out of other markets or venues where borrowing could have led to alternative negotiations, terms, or even trade. Debts accumulated by an oppressive regime are automatically suspect, but what about when its subject, a colony lacking in sovereignty and economic decision-making authority, resorts to borrowing in order to survive? Should we presume it legitimate? Despite the prevailing narrative that Puerto Rico should pay what it owes, who in fact is the responsible party? If Puerto Rico cannot exercise control over its economic future or negotiate with full autonomy, how can there be a fair and informed negotiation?

Puerto Rico cannot renegotiate its debt alone, nor can it seek other sources of financing or resort to mechanisms in regional and international forums that could have offered it better terms and conditions. On top of that, the US Supreme Court affirmed Puerto Rico's colonial status in 2016, days before the passage of PROMESA, giving Congress carte blanche under its plenary powers to draft legislation that would only deepen the political subservience of the island to the United States' economic interests.[3] As a result, it is the Fiscal Control Board that now governs Puerto Rico, depriving Puerto Ricans even further of their civil and political rights. With the board's imposition of shattering austerity measures to further its stated goal of debt repayment, Puerto Ricans' social, cultural, and economic rights continue to suffer without end, despite calls by the United Nations to give up such misguided policies.

Puerto Rico's debt is odious because it perpetuates colonialism. It does so by financing colonial regimes and propping them up for the economic extraction in service to other nations or financial institutions. In this historical and political moment of our global society, colonialism and the colonial legacies of debt they create must be widely repudiated, along with the mechanisms and structures that finance them in order to profit from their perpetual political and economic crisis.

1. That case was appealed up to the US Supreme Court, where, in a decision written by Justice Clarence Thomas, the law was indeed found to be unconstitutional. See Puerto Rico v. Franklin California Tax-Free Tr., 136 S. Ct. 1938, 195 L. Ed. 2d 298 (2016).

2. "Puerto Rico Debt Crisis: 'Human Rights Cannot Be Sidelined'—UN Expert Warns," United Nations Human Rights Office of the High Commissioner, January 9, 2017.

3. Puerto Rico v. Sanchez Valle, 136 S. Ct. 1863, 195 L. Ed. 2d 179 (2016).

DISMANTLING PUBLIC EDUCATION IN PUERTO RICO[1]

Rima Brusi and Isar Godreau

Hurricane María hit Puerto Rico's public education system hard. The island's K–12 schools suffered $142 million in damages, and the campuses of the University of Puerto Rico (UPR), $133 million.[2] Shortly after the storm, students, teachers, family members, and administrators who could drive rushed to help clear away the debris from fallen ceilings and broken windows, and rescue libraries and equipment from water damage. María's effects, however, went beyond this immediate material impact. The storm also provided a convenient excuse for accelerating and intensifying the process of shrinking and weakening the public sector in ways that benefit private and corporate interests, a process that had started well before the storm appeared in weather forecasts and news reports.

Two years before the hurricane, in 2015, Puerto Rico caught the attention of the *New York Times* and other media outlets when then governor Alejandro García Padilla publicly declared the island's $72 billion bond debt "unpayable" and announced that his government would seek "significant concessions" from debt holders.[3] Shortly thereafter, a report analyzing the crisis was commissioned by the Ad Hoc Group of Puerto Rico's General Obligation Bondholders—composed mostly of hedge funds that had purchased high-risk, high-interest debt on the cheap and that had lobbied heavily against any form of bankruptcy or relief for the territory.[4] The report highlighted a series of recommendations for Puerto Rico's government that were explicitly built on two assumptions: (1) that Puerto Rico's financial woes are "fixable" because they stem not from a "debt problem" proper but from financial mismanagement, and (2) that the governor's proposal to extract concessions from debtors posed significant legal and financial risks, and that therefore Puerto Rico needed to focus on fully paying its debt. The report, authored by three people with ties to the International Monetary Fund and provocatively entitled *For Puerto Rico, There Is a Better Way*, includes four recommendations for "fiscal reform measures," two of which target the public education system. One proposal called for reducing the number of teachers in the K–12 system, the other for significantly reducing funding for the University of Puerto Rico (from here on UPR), which the authors refer to as a "subsidy."

The following year, two major events dramatically shaped Puerto Rico's political and fiscal situation: García Padilla lost the general election to pro-statehood candidate Ricardo Rosselló, who ran on a platform that included a promise to pay bondholders, and the US Congress passed the PROMESA Law, which established a Fiscal Oversight and Management Board, a non-elected body of seven members locally referred to as la Junta.[5] La Junta has nearly absolute powers over Puerto Rico's finances and can overturn local laws that may interfere with the implementation of fiscal austerity measures. Soon after their first official meeting in Wall Street, Junta members targeted both the public K–12 and higher education systems of Puerto Rico in ways that closely followed the bondholders' "recommendations" in the *Better Way* report. Correspondingly, Governor Rosselló formally announced that the school closures already underway from previous administrations were going to intensify, with almost half the island's schools now targeted for closing. Following suit, the Financial Oversight and Management Board (FOB) demanded that the university cut about a third of its overall budget.[6] And so, at the start of 2017, before Hurricane María made landfall on the southeastern town of Yabucoa on September 20, Puerto Rico's public K–12 schools and eleven university campuses were already facing an uncertain future. By the time the hurricane dealt a devastating blow to the island's public education infrastructure, the system was already under siege from economic policies that disinvested in infrastructure, equipment, and teaching personnel. After the hurricane, disaster relief efforts did not slow down this process; rather, they accelerated the dismantling of Puerto Rico's schools and public university.

"REFORMING" K–12: *LA REFORMA EDUCATIVA*

The process of progressively defunding public education before and after the storm required highly paid technocrats and bureaucrats committed to implementing the "difficult" decisions recommended by la Junta and by PROMESA stakeholders. Rosselló had tapped Julia Keleher for the position of secretary of education in late 2016. In the four years leading up to her appointment, Keleher's education consultancy firm, Keleher and Associates, had been awarded almost $1 million in contracts to "design and implement education reform initiatives" in Puerto Rico.[7] The results of those efforts were never described to the public, but her salary as secretary, double the size of her predecessors', was justified according to her "world-class skills" and credentials.[8]

After becoming secretary, and especially after the hurricane, Keleher and her department not only accelerated the pace of school closures, citing hurricane-triggered migration, but also, with the assistance of US Education Secretary Betsy DeVos's office, produced an education-reform bill explicitly designed to increase "school choice" through charter schools and school-voucher programs.[9] Critics and activists quickly voiced their opposition. In the Senate hearings leading up to the bill's approval, UPR's dean of education testified against the widespread adoption of charter schools using research-based arguments. He was removed from his post shortly after.[10] The bill was signed into law by Rosselló a few months after Hurricane María, in March 2018.[11] Tellingly, a separate law enacted around the same time allows for the fast-tracking of so-called church schools, celebrated by their advocates as a way to bring together "religious freedom" and "school choice."[12] One of these schools, formerly named after one of Puerto Rico's national poets, Julia de Burgos, was rented by the government to an evangelical church for one dollar a month.[13] The church turned it into a private school, renamed it Fountain Christian Bilingual School, and, as part of the building's remodeling, painted over and effectively destroyed a 1966 mural by the renowned Puerto Rican artist José Torres Martinó.

The entire incident—privatizing a public school, changing its name from a Puerto Rican poet's to a religious name in English, and painting over a valuable artwork—is an apt metaphor for the transformation of the system and evidence of Keleher's self-admitted cultural incompetence.[14] In fact, in a move reminiscent of the actions of unelected colonial governments of the early twentieth century (e.g., replacing Spanish with English as the language of instruction and punishing displays of the Puerto Rican flag on school grounds), the secretary also eliminated Puerto Rican Week from the department's official curriculum.[15] She also recruited candidates for high-level leadership positions from the fifty states instead of the island.[16]

Government officials refer to the new law as *la reforma educativa* (education reform). The name is fitting; when in the 1990s Rosselló's father, Pedro Rosselló, was governor, he privatized the public health system, and both the new health system and the law that created it were called *la reforma de salud* (health reform). It is hard not to read today's education law, which consists mainly of eliminating public schools and creating charter schools and voucher systems, as anything other than another large-scale privatization of the education system. The final bill is, moreover, frustratingly vague on some key issues: charters may or may not be for-profit, vouchers may or may not be used at religious schools, and charters and private schools may or may not accept special-education students.

This last point, about students with special needs, is crucial. Unlike in the fifty states, where about 13 percent of the student body qualifies for special-education services, in Puerto Rico 40 percent of the student population requires them.[17] But research suggests that charter schools are less likely than traditional public schools to enroll and retain students with disabilities.[18]

According to Keleher and other government officials, the school closures are justified because the student population was greatly reduced after the hurricane and because the closed schools were not teaching their students effectively. Yet the number of students had not dropped significantly at some of the schools that were closed, and, as critics argue, closing neighborhood schools can become a cause rather than a result of migration, since many families live below the poverty line and lack adequate transportation.[19] Some closed schools, moreover, were actually considered "excellent" by the department itself, and some of the schools receiving new students as a result of the closings lack the necessary facilities to accommodate an increased student population.[20] This led to the use of containers (provided by a private contractor paid with FEMA hurricane relief funds) to house students, which effectively excludes some students with disabilities who require special accommodations and the implementation of shortened or "interlocking" schedules that shorten students' academic activities.

HIGHER ED "RIGHTSIZING": *REFORMA UNIVERSITARIA*

A similar turn of events unfolded with public higher education. Before Hurricane María dealt a blow to its infrastructure in 2017, the UPR system was already facing the draconian budget cuts imposed by la Junta. Members of la Junta were making public declarations about the need for huge cuts to the public university's budget as early as January 2017. The actual numbers were a moving target: first $350 million, then $450 million, then $500 million.[21] The rationale for each calculation was never made public, but the cuts represented about a third of the system's total budget. After more than a year of ignoring feedback and alternative fiscal plans drafted by the university leadership as well as student and faculty groups, la Junta imposed its own plan in April 2018, which immediately doubled tuition, with increases of up to 175 percent soon to come.[22] The plan also unveiled a euphemistically described "campus consolidation" that will likely close or shrink seven campuses and dramatically reduce the student body, faculty, and staff.[23] Ironically, Puerto Rico's government is forced, by law, to cover la Junta's expenses, which amounted to $31 million in just the first ten months and are projected to reach well over $300 million over five years.[24]

After the hurricane, rather than rushing to repair and strengthen this crucial asset for Puerto Rico's socioeconomic development, the federal and local governments seemed intent on further damaging the island's public university system. In a territory with soaring poverty and unemployment rates, la Junta has not only doubled tuition but also eliminated the waivers traditionally offered to athletes, choir members, and other students providing services to the university—all at an institution that for over a century has been the island's main channel for upward mobility.[25]

The size, scope, and pace of the budget cuts demanded by la Junta increased after the hurricane and were euphemistically presented as a "rightsizing" of nearly $550 million over the next five years.[26] The federal government did little to mitigate the difficult situation produced by the combined effect of the hurricane and the augmented austerity measures advocated by la Junta. In the weeks after Hurricane María, the US Department of Education made $41 million available to support students at colleges and universities impacted by the hurricane, of which the UPR system received only 20 percent.[27] For comparison, in 2005 institutions in Louisiana and Mississippi were able to access $190 million after Hurricane Katrina. Advocates criticized the design of the application process and its forms, which made it onerous, even impossible, for UPR campuses to apply for most funding in the midst of power outages and rebuilding efforts.[28] Adding insult to injury, a considerable portion of the María relief funds were awarded to institutions that had not been directly affected by the storm, including private ones like New York University and even some for-profit ones like Grand Canyon University, to host a relatively small number of students who came from Puerto Rico and other territories or states affected by hurricanes Irma or María.

In addition to making few funds available to rebuild UPR, the federal government has also gutted essential aid programs, like work-study grants, with little explanation. In Puerto Rico, where the median annual household income is below $20,000, these policies are particularly damaging for low-income students who will likely have to drop courses in order to work part-time and secure a livable income.[29]

La Junta has offered no rationale for why, from the very beginning, it targeted UPR with such drastic cuts. It is an odd decision

240 | *Aftershocks of Disaster*

considering that the UPR is one of the strongest contributors to upward mobility and the local economy and one of the public entities that has historically best managed its own debt.[30] Such a relentless handicapping of the public university's function and mission would seem to make little economic sense for Puerto Rico at this historical juncture. One would expect the government to facilitate and encourage research initiatives in all areas after Hurricane María, from the creation of new solar technologies to the treatment of social trauma. The UPR is a crucial center for research, generating over 70 percent of the scientific research output on the island.[31] Despite carrying heavy teaching loads with few of the resources that faculty in the continental United States often take for granted, UPR has a world-class faculty that includes award-winning humanists and scientists, and has been an important site of scientific innovation and critical thinking. While many universities in the states struggle to increase the number of STEM degrees they produce, particularly among Latinx students, the UPR is one of the top schools in the country graduating STEM students at the baccalaureate and graduate levels.[32] The UPR owns and runs public hospitals, museums, theaters, and public libraries. It prepares the best K–12 teachers on the island.[33] It is a land-grant and sea-grant institution, and as such, it owns and operates botanical gardens, offers agricultural-extension services, and runs programs to help protect and responsibly utilize coastal areas.

Given the depth and range of the university's positive impact on Puerto Rico's economy and society and PROMESA's expressed goal of promoting economic development for Puerto Rico, one would expect that la Junta would make the UPR the last, not the first, target for austerity measures. Faculty and students on the ground have some ideas as to why it was just the opposite. Austerity, not economic development, has been la Junta's focus. La Junta's neoliberal rationale is that moving investments from public to private hands will have a trickle-down effect on the economy. The public university, however, has historically been a site of resistance to such neoliberal reforms, whose failures have been openly questioned and described by faculty experts, students, and workers. Indeed, higher education is, generally speaking, strongly linked to increased political participation, and a recent study carried out in Puerto Rico by one of the coeditors of this book, Yarimar Bonilla, found that those who attended public institutions had much higher rates of what social scientists describe as "political knowledge" than those who attended private institutions.[34] UPR's role as a cradle of island-wide political movements thus poses a threat to the highly unpopular Junta. Moreover, UPR is known for a robust student movement that has vigorously protested colonial intervention since 1948 and spearheaded national resistance to austerity measures in 1984, 2010, and, most recently, against la Junta itself in 2017.[35] Weakening UPR and its potential for resistance would no doubt favor a government body intent on slashing public resources.

Furthermore, the depth and range of the university's positive impact on Puerto Rico's economy and society also challenges an agenda of austerity and disaster profiteering that, in stifling economic development, has no interest in developing Puerto Rico's youngest and brightest. With a plan that does not promote economic growth but rather focuses on austerity and the transferring of public funds to private profit, it would be a liability to have a highly educated group among the ranks of the unemployed or minimum-wage workers.

FROM PUBLIC GOOD TO PRIVATE PROFIT

The attacks on UPR and the public K–12 system cannot be viewed in isolation. We must take into account that, right after Hurricane María, private, non–Puerto Rican firms were given large contracts to carry out the process of rebuilding—often with disastrous results.[36] Now plans are underway to sell Puerto Rico's electric utility as the government invites hedge funds and bankers associated with the blockchain industry to invest in Puerto Rico.[37] Protected areas of high ecological value are also being targeted for sale, and a portion of the public-education budget will be funneled into the coffers of private and charter schools.[38] These transformations need to be understood as part of a broad, violent takeover of disaster capitalism, in which those who stand to profit from Puerto Rico's tragedy have been gifted by not one but two disasters: Puerto Rico's unpayable debt (around $72 billion in bonds and $50 billion in pension obligations), and Hurricanes

Irma and María, with estimated damages of $90 billion ($133 million for the university) and a death toll in the thousands.[39]

How will disaster capitalism play out for public schools and college campuses? Some of the implications are quite predictable. On an island where over 40 percent of the population and over 50 percent of children live under the poverty line, the students affected most by these cuts are precisely the ones with the greatest need.[40] This is not only because of the potential exclusion of special education students from charter schools or a spiked university tuition, but also because some of the closed schools and smaller campuses targeted for closure largely serve low-income, place-bound students in poorer municipalities. These students are likely to drop out, migrate, or be absorbed by the US mainland's charter or for-profit college industry, which already markets heavily to them—with some of the very same actors and substandard results.

In higher education, for-profit companies are, unsurprisingly, lobbying heavily and successfully for decreased regulation and more freedom from public scrutiny, and institutions serving Puerto Rican students are no exception.[41] A large, nominally nonprofit system on the island is expanding its for-profit arm, targeting Puerto Rican students who have moved to Florida, and for-profit colleges are vying for a contract to take over Puerto Rico's police academy.[42] In this context, it is very telling that Puerto Rico's only (and nonvoting) representative in the House, Resident Commissioner Jennifer González, recently submitted a bill designed to relax the regulations on for-profit higher-education institutions in Puerto Rico.[43] Specifically, the bill seeks to increase these colleges' access to public funding. González has not used her position in Washington to advocate for the public university, even though it graduates the most students on the island, has a better record of gainful employment for alumni than its private counterparts, and has a better and greater impact than them on the economy.

Most ironically, some of the actors that stand to benefit from the privatization of Puerto Rico's educational institutions are closely connected to holders of Puerto Rico's bond debt. This is the case, for example, of Apollo Education Group, the corporation behind the

University of Phoenix, which is partly owned by Apollo Global Management, a bondholder in Puerto Rico's debt.[44] Similarly, Canyon Capital and Stone Lion Capital, two of the main holders of Puerto Rico's bond debt, have strong links with the charter school industry.[45]

The dangers faced by students at for-profits in Puerto Rico are the same as those that have been described in the fifty states. With few exceptions, for-profits have dismal graduation rates and offer low-quality degrees. Students who attend them are often left in debt with no degree and no job.[46] UPR, by contrast, has the opposite track record: it boasts the best overall graduation rate in Puerto Rico and has been recognized by many scholars as an engine of social mobility that helped move previous generations out of poverty and into middle-class status. While student advocates in the United States denounce the fact that only about 10 percent of flagship public colleges are affordable for low-income students, 64 percent of students at UPR's main campus are low income.

Renowned economists around the world, and even the IMF itself, have concluded that we need more, not fewer, of these examples.[47] Austerity measures cannot help broken economies and in fact do more harm than good. The attacks on Puerto Rican public education and economic development as a whole will further decrease upward mobility, hamper economic growth, and increase the island's already soaring social inequality. Indeed, austerity measures combined with failure to promote economic development before Hurricane María have made Puerto Rico one of the countries with the highest Gini coefficient, a standard measure of economic inequality, ranking fifth in the world. After the hurricane social inequality worsened, and Puerto Rico ranked third among 101 countries overall, surpassed only by Zambia and South Africa.[48]

Cutting funding to the public education system spurs the growth of private and for-profit institutions and reduces access to affordable educational opportunities, exacerbating the negative effects of María and increasing the island's already out of control inequality. In this context, Puerto Rico's colonial condition and its concomitant lack of sovereign powers facilitate not only the payment of a dubious debt to the detriment of social services but also profit-making for powerful and extraneous economic actors, some of which are also debt holders, all at the expense of people's right to access quality education and improve their life chances. Marginalizing those who raise their voices against these failed strategies and defunding the public institutions that promote this awareness might quell political resistance to austerity temporarily, in the aftershocks of the hurricane. But the rush to implement this failed script in Puerto Rico, and to let it guide policy, especially in the delicate and important realm of public education, will surely lead to more inequality, increased unrest, and resistance. For students, their struggling families, and other vulnerable groups, this man-made storm started long ago and shows no signs of stopping.

1. Some of the content of this chapter has previously appeared in the *Nation* and *80 Grados*.

2. Kyra Gurney, "Kids Are Back in School in Puerto Rico. But Hurricane María's Effects Still Linger," *Tampa Bay Times*, August 14, 2018, https://www.tampabay.com/news/education/Kids-are-back-in-school-in-Puerto-Rico-But-Hurricane-Maria-s-effects-still-linger_170882206; Claire Cleveland, "Without Researchers or Funds, Puerto Rico Universities Grapple with Future after Hurricane María," *Cronkite News*, May 4, 2018, https://cronkitenews.azpbs.org/2018/05/04/puerto-rico-universities-grapple-with-future-after-hurricane-Maria/.

3. Michael Corkery and Mary Williams Walsh, "Puerto Rico's Governor Says Island's Debts Are 'Not Payable,'" *New York Times*, June 28, 2015, https://www.nytimes.com/2015/06/29/business/dealbook/puerto-ricos-governor-says-islands-debts-are-not-payable.html.

4. Sheeraz Raza, "For Puerto Rico, There Is a Better Way," *ValueWalk*, July 27, 2015, https://www.valuewalk.com/2015/07/for-puerto-rico-there-is-a-better-way/.

5. Joanisabel González, "Ricardo Rosselló buscará pagar la deuda del País," *El Nuevo Día*, August 14, 2016, https://www.elnuevodia.com/negocios/economia/nota/ricardorossellobuscarapagarladeudadelpais-2230317/; Susan Cornwell and Nick Brown, "Puerto Rico Oversight Board Appointed," Reuters, August 31, 2016, https://www.reuters.com/article/us-puertorico-debt-board-idUSKCN11628X.

6. Rima Brusi, Yarimar Bonilla, and Isar Godreau, "When Disaster Capitalism Comes for the University of Puerto Rico," *Nation*, September 20, 2018, https://www.thenation.com/article/when-disaster-capitalism-comes-for-the-university-of-puerto-rico/.

7. Kelia López Alicea, "A Defense for Keleher's Contract," *El Nuevo Día*, February 15, 2017, https://www.elnuevodia.com/english/english/nota/adefenseforkeleherscontract-2291546/.

8. Metro Puerto Rico, "Rosselló afirma que Julia Keleher es una profesional de calibre 'global,'" March 8, 2018, https://www.metro.pr/pr/noticias/2018/03/08/rossello-afirma-julia-keleher-una-profesional-calibre-global.html.

9. Ley de Reforma Educativa de Puerto Rico, PR.gov, March 29, 2018, http://www2.pr.gov/ogp/BVirtual/LeyesOrganicas/pdf/85–2018.pdf.

10. Laura M. Quintero, "Sacan al decano de la Facultad de Educación," El Vocero, October 3, 2018, https://www.elvocero.com/educacion/sacan-al-decano-de-la-facultad-de-educaci-n/article_988c10f8-2405-11e8-a9b9-a3a33b30498a.html.

11. Daniel Rivera Vargas, "Rosselló convierte en ley la reforma educativa," Primera Hora, March 29, 2018, https://www.primerahora.com/noticias/gobierno-politica/nota/rosselloconvierteenleylareformaeducativa-1275122/.

12. Índice, "Firma ley que exime Iglesias Escuela de regulación estatal," June 7, 2017, http://www.indicepr.com/noticias/2017/06/07/news/70913/firma-ley-que-exime-iglesias-escuela-de-regulacion-estatal/.

13. NotiCel, "Nueva escuela de Font 'destruye' obra de arte," April 12, 2018, https://www.noticel.com/ahora/educacion/nueva-escuela-de-font-borra-obra-de-arte/728692114.

14. NotiCel, "Keleher discute con maestros durante taller de capacitación," March 9, 2018, https://www.noticel.com/ahora/educacion/keleher-discute-con-maestros-durante-taller-de-capacitacin/713826409.

15. Ayala César and Rafael Bernabe, *Puerto Rico in the American Century* (Chapel Hill: University of North Carolina Press, 2007); Félix Cruz, "Keleher elimina la Semana de la Puertorriqueñidad," El Post Antillano, August 2, 2017, http://elpostantillano.net/cultura/19903-2017-08-02-17-08-12.html.

16. There is much scholarship on this topic. See, for example, Jorge R. Schmidt, *The Politics of English in Puerto Rico's Public Schools* (Boulder, CO: Lynne Rienner, 2014).

17. National Center for Education Statistics, "Children and Youth with Disabilities," April 2018, https://nces.ed.gov/programs/coe/indicator_cgg.asp; Agencia EFE, "Se cuadruplican estudiantes de educación especial en Puerto Rico en 10 años," Primera Hora, December 15, 2015, https://www.primerahora.com/noticias/mundo/nota/secuadruplicanestudiantesdeeducacionespecialenpuertoricoen10anos-1126676/.

18. Mary Bailey Estes, "Choice for All? Charter Schools and Students with Special Needs," *Journal of Special Education* 37 (2004): 257–67.

19. TeleSur, "¿Por que el Gobierno de Puerto Rico cerrará 300 escuelas?," February 16, 2018, https://www.telesurtv.net/news/cierran-escuelas-puerto-rico-20180216-0040.html; Nydia Bauza, "Fuerte oposición a la clausura de escuelas," Primera Hora, April 8, 2018, https://www.primerahora.com/noticias/gobierno-politica/nota/fuerteoposicionalaclausuradeescuelas-1276720/.

20. Laura M. Quintero, "En lista de cierre 56 escuelas de excelencia," *El Vocero*, April 14, 2018; El Nuevo Día, "Educación comienza a instalar vagones en los que se dara clases en las escuelas," August 8, 2018, https://www.elnuevodia.com/noticias/locales/nota/educacioncomienzaainstalarvagonesenlosquesedaraclasesenlasescuelas-2440195/.

21. Juan Giusti Cordero, "El misterio de los $450 + millones y la UPR," *80 Grados*, June 23, 2017, https://www.80grados.net/el-misterio-de-los-450-millones-y-la-upr/.

22. Sin Comillas, "Profesores del RUM presentan un Plan Fiscal sostenible para la UPR," March 5, 2018, http://sincomillas.com/profesores-del-rum-presentan-un-plan-fiscal-sostenible-para-la-upr/; Kelia López Alicea, "Fiscal Blow to the UPR," *El Nuevo Día*, April 25, 2018, https://www.elnuevodia.com/english/english/nota/fiscalblowtotheupr-2417518/.

23. University of Puerto Rico, "New Fiscal Plan for University of Puerto Rico," October 21, 2018, http://www.upr.edu/wp-content/uploads/2018/10/Fiscal-Plan-21-oct-2018-.pdf.

24. Univision, "Junta de Supervision Fiscal gastó casi 31 millones de dólares en 10 meses," August 1, 2017, https://www.univision.com/puerto-rico/wlii/noticias/junta-de-control-fiscal/junta-de-supervision-fiscal-gasto-casi-31-millones-de-dolares-en-10-meses; Jose A. Delgado, "Junta Control Fiscal costará cientos de millones de dólares," *El Nuevo Día*, June 4, 2016, https://www.elnuevodia.com/noticias/politica/nota/juntacontrolfiscalcostaracientosdemillonesdedolares-2206623/.

25. Walter Díaz, "Universidad y Capital Humano: Clase Social y Logro Educativo en Puerto Rico," Centro Universitario para el Acceso, University of Puerto Rico at Mayagüez, April (2010).

26. Keria López Alicea, "La Junta recorta 10% de los gastos de la UPR," *El Nuevo Día*, April 25, 2018, https://www.elnuevodia.com/noticias/locales/nota/lajuntarecorta10delosgastosdelaupr-2417425/.

27. Erica L. Green and Emily Cochrane, "In Devastated Puerto Rico, Universities Get Just a Fraction of Storm Aid," *New York Times*, May 1, 2018, https://www.nytimes.com/2018/05/01/us/politics/hurricane-maria-puerto-rico-emergency-aid.html.

28. Rafael Medina, "Release: CAP Submits Comments Denouncing DeVos for Ripping Off Puerto Rico's IHEs; Calls on DeVos to Abandon the Form-Based Process and Provide Guidance in Spanish," Center for American Progress, April 6, 2018, https://www.americanprogress.org/press/release/2018/04/06/449155/release-cap-submits-comments-denouncing-devos-ripping-off-puerto-ricos-ihes-calls-devos-abandon-form-based-process-provide-guidance-spanish/.

29. United States Census Bureau, "Population Estimates, July 1, 2018 (V2018)," https://www.census.gov/quickfacts/pr.

30. Eduardo Berrios Torres, "La verdad sobre el Plan de Retiro de la UPR," *El Nuevo Día*, September 11, 2018, https://www.elnuevodia.com/opinion/columnas/laverdadsobreelplanderetirodelaupr-columna-2446483/.

31. Jose I. Alameda-Lozada and Alfredo González-Martínez, "El impacto socioeconómico del sistema de la Universidad de Puerto Rico," *Occasional Papers*, no. 7 (2017): http://www.estudiostecnicos.com/pdf/occasionalpapers/2017/OP-No-7-2017.pdf.

32. Kimberly Leonard, "Building a Latino Wave in STEM," US News and World Report, May, 19, 2016, https://www.usnews.com/news/articles/2016-05-19/building-a-latino-wave-in-stem; Deborah A. Santiago, "Finding your Workforce: The Top 25 Institutions Graduating Latinos," Excelencia in Education, May 2012, https://www.edexcelencia.org/research/publications/finding-your-workforce-top-25-institutions-graduating-latinos.

33. María de los Angeles Ortiz, "Informe final presentado al Consejo de Educación Superior de PR sobre indicadores de calidad en los programas de preparación de maestros en cuatro IES en Puerto Rico," *Division de Investigacion y Documentacion sobre la Educación Superior del Consejo de Educación Superior de Puerto Rico* (Oritz, Lord, Hope & Associates, 2006); "Ejemplares 8 programas de Preparación de Maestros de la UPR," last modified January 18,

2017, https://www.metro.pr/pr/noticias/2017/01/18/ejemplares-8-programas
-preparacion-maestros-upr.html.

34. D. Sunshine Hillygus, "The Missing Link: Exploring the Relationship
between Higher Education and Political Engagement," *Political Behaviour*
27, no. 1 (2005); Jim Patterson, "Education Is the Key to Promoting Political
Participation: Vanderbilt Poll," Research News @Vanderbilt, June 25, 2012,
https://news.vanderbilt.edu/2012/06/25/education-key-to-promoting
-political-participation/.

35. Juan Carlos Castillo, "Las huelgas estudiantiles de la UPR, aquellas que se re-
piten y continúan (Parte 1)," *Diálogo UPR*, June 30, 2015, http://dialogoupr
.com/las-huelgas-estudiantiles-de-la-upr-aquellas-que-se-repiten
-y-continuan-parte-i/; Rima Brusi-Gil de Lamadrid, "The University of
Puerto Rico: A Testing Ground for the Neoliberal State," *NACLA Report on
the Americas*, May 12, 2011, https://nacla.org/article/university-puerto-rico
-testing-ground-neoliberal-state; Cynthia López Cabán, "La mayoría de los
recintos de la UPR están en huelga indefinida," *El Nuevo Día*, April 6, 2017,
https://www.elnuevodia.com/noticias/locales/nota
/lamayoriadelosrecintosdelauprestanenhuelgaindefinida-2307616/.

36. Vann R. Newkirk II, "The Puerto Rican Power Scandal Expands," *Atlantic*,
November 3, 2017, https://www.theatlantic.com/politics/archive/2017/11
/puerto-rico-whitefish-cobra-fema-contracts/544892/; Patricia Mazzei and
Agustin Armendariz, "FEMA Contract Called for 30 Million Meals for
Puerto Ricans. 50,000 Were Delivered," *New York Times*, February 6, 2018,
https://www.nytimes.com/2018/02/06/us/fema-contract-puerto-rico.html.

37. Dawn Giel and Seema Mody, "Puerto Rico Lures Blockchain Industry to
Help Fund Its Comeback," *CNBC*, March 16, 2018, https://www.cnbc.com
/2018/03/16/puerto-rico-lures-blockchain-industry-to-help-fund-its
-comeback.html.

38. Gerardo E. Alvarado Leon, "El gobierno mercadea 17 terrenos protegidos,"
El Nuevo Día, May 10, 2018, https://www.elnuevodia.com/noticias/locales
/nota/elgobiernomercadea17terrenosprotegidos-2421335/.

39. Daniel Uria, "Hurricane María Caused $90B of Damage in Puerto Rico," UPI,
April 9, 2018, https://www.upi.com/Hurricane-Maria-caused-90B-of
-damage-in-Puerto-Rico/6421523309427/; Claire Cleveland, "Without Researchers or Funds, Puerto Rico Universities Grapple with Future after Hurricane María," *Cronkite News*, May 4, 2018, https://cronkitenews.azpbs.org/2018
/05/04/puerto-rico-universities-grapple-with-future-after-hurricane-Maria/.

40. United States Census Bureau, "Population Estimates, July 1, 2018 (V2018),"
https://www.census.gov/quickfacts/pr; Bianca Faccio, "Left Behind: Poverty's Toll on the Children of Puerto Rico," *Child Trends*, March 28, 2016,
https://www.childtrends.org/left-behind-povertys-toll-on-the-children
-of-puerto-rico; Rima Brusi, Walter Díaz, and David González, "So Close
and So Far: 'Merit,' Poverty, and Public Higher Education in Puerto Rico,"
Revista de Ciencias Sociales, 2010, https://revistas.upr.edu/index.php/rcs
/article/view/7470/6074.

41. Michelle Chen, "Why Are New York's Public Funds Going to For-Profit College Tuition?," *Nation*, April 5, 2018, https://www.thenation.com/article/why-are-new-yorks-public-funds-going-to-for-profit-college-tuition/.
42. Brook v. Sistema Universitario Ana G. Mendez, Inc., 8.17 (M.D. Fla. Nov. 20, 2017); Keila López Alicea, "Una universidad entrenaria a los futuros policias," *El Nuevo Día*, July 6, 2018, https://www.elnuevodia.com/noticias/seguridad/nota/unauniversidadentrenariaalosfuturospolicias-2433190/.
43. US Congress, House, *Puerto Rico Higher Education Disaster Relief Act*, HR 5850, 115th Congress, https://www.congress.gov/bill/115th-congress/house-bill/5850/text?format=txt.
44. Alan Mintz et al., "Hedgepapers No. 17—Hedge Fund Vultures in Puerto Rico," Hedge Clippers, July 10, 2015, http://hedgeclippers.org/hedgepapers-no-17-hedge-fund-billionaires-in-puerto-rico/.
45. Joel Cintron Arbasetti, "La trayectoria de los fondos de cobertura que llegaron a Puerto Rico," Centro de Periodismo Investigativo, July 14, 2015, http://periodismoinvestigativo.com/2015/07/la-trayectoria-de-los-fondos-de-cobertura-que-llegaron-a-puerto-rico/.
46. Mamie Lynch, Jennifer Engle, and Jose Luis Cruz, "Subprime Opportunity: The Unfilled Promise of For-Profit Colleges and Universities," Education Trust, November 22, 2010, https://edtrust.org/resource/subprime-opportunity-the-unfulfilled-promise-of-for-profit-colleges-and-universities/.
47. Yalixa Rivera and Jonathan Levin, "Puerto Rico's Fiscal Plan Was Doomed Even before María, Stiglitz Says," *Bloomberg*, January 16, 2018, https://www.bloomberg.com/news/articles/2018-01-16/puerto-rico-fiscal-plan-doomed-even-before-maria-stiglitz-says; Larry Elliott, "Austerity Policies Do More Harm Than Good, IMF Study Concludes," *Guardian*, May 27, 2016, https://www.theguardian.com/business/2016/may/27/austerity-policies-do-more-harm-than-good-imf-study-concludes.
48. *El Nuevo Día*, "Puerto Rico es el tercer país de mayor desigualdad económica del mundo," September 17, 2018, https://www.elnuevodia.com/negocios/economia/nota/puertoricoeseltercerpaisdemayordesigualdadeconomicaenelmundo-2447734/.

PUERTO RICO'S FIGHT FOR A CITIZEN DEBT AUDIT

A Strategy for Public Mobilization and a Fair Reconstruction

Eva L. Prados-Rodríguez[1]

Puerto Rico is in pain. It is experiencing the combined force of a 120-year colonial relationship with the United States and a fiscal and economic crisis that has led it to accumulate over $74 billion in odious, illegitimate, unsustainable, and potentially illegal public debt that exceeds 100 percent of our gross national product (GNP). We are also living through a humanitarian crisis caused by the devastation of hurricanes Irma and María, and the gross negligence of the federal and local response to both events.

The effects of these combined forces have been felt, for example, in the imposition of the unelected Fiscal Control Board, in 2016, to restructure Puerto Rico's public debt, dismantle its public sphere, and advance an aggressive privatization and austerity agenda.[2] Also in the thousands of people who died as a result of the country's unpreparedness for the events of September 20, 2017.[3]

To cope with these accumulating and intersecting crises, more than 150,000 residents of Puerto Rico have organized behind the banners of transparency and accountability,[4] and have created the Citizens' Front to Audit the Debt (Frente Ciudadano por la Auditoría de la Deuda): a nonpartisan and wide-ranging coalition in Puerto Rico consisting of grassroots organizations, professional associations, labor unions, businesspeople, academics, students, human rights defenders, and other citizens who advocate for a comprehensive audit of Puerto Rico's public debt. When the Puerto Rican government eliminated a public commission created to conduct a public audit of the debt, we created the Citizens' Commission for the Comprehensive Audit of the Public Debt, a nonprofit organization whose only purpose is to perform a transparent and comprehensive citizen audit of Puerto Rico's debt issuances for the past five decades.[5]

A comprehensive citizen debt audit would allow us to evaluate (1) how much is really owed in principal and interest, (2) the legality and legitimacy of each bond issuance, (3) the conduct of every involved actor and financial institution, (4) how the money was spent, and (5) how political and economic elements, such as colonialism and financial capitalism, played a crucial role in the accumulation of Puerto Rico's public debt.[6] Such an audit would also become a powerful tool toward fostering citizen involvement, participation, and mobilization in the decision-making processes related to Puerto Rico's public debt and the policy measures advanced by the government and the Fiscal Control Board to address our fiscal and economic crisis.[7]

Our collective insists that it is unsustainable for the people of Puerto Rico to pay $74 billion in debt without a complete, independent, and formal audit. While austerity measures have been implemented in Puerto Rico for decades, recent Fiscal Control Board intervention on behalf of unaudited creditors has pushed for more severe budget cuts, privatizations, and the elimination and/or reduction of workers' benefits and pension plans.[8]

Puerto Rico Law 97 of 2015 created the Commission for the Comprehensive Audit of Puerto Rico's Public Debt, an independent public entity with the stated purpose of auditing all components of Puerto Rico's currently outstanding debt in accordance with government accounting and legal standards for performance and compliance, among other standards, as needed. Unfortunately, the government dragged its feet on performing such a debt audit, first delaying the appointment of its citizen representatives, then severely underfunding the commission,[9] until the administration of Governor Ricardo Rosselló finally disbanded the commission[10] after unsuccessfully attempting to dismiss its citizen representatives.[11]

Through its preaudit reports, this commission, as well as studies published by other organizations, uncovered a lot of issues regarding the legality of Puerto Rico's public debt. For example, a preaudit report published by the commission argued that about half of Puerto Rico's total debt is potentially illegal and in violation of our Constitution.[12]

Additionally, other published reports identified several predatory practices by Puerto Rico's creditors. For instance, Puerto Rico has $37.8 billion in outstanding capital appreciation bonds (CAB), accounting for a very large share of its total outstanding debt. A CAB is a long-term bond with compound interest on which the borrower does not make any principal or interest payments for the first several years and, in some cases, until the final maturity of the bond. The principal on these bonds is just $4.3 billion. The remaining $33.5 billion is interest—that means we have loans on which we are supposed to pay eight or nine times the amount we originally borrowed.[13] Hence, $33.5 billion of the island's outstanding debt isn't debt at all but unpaid interest. This is predatory and illegitimate debt, and it should not be paid.

Many of the investors who now own Puerto Rico's debt, including vulture hedge funds, never expected the island to be able to repay all of it because it had already been written down as bad debt—which is what allowed these investors to buy it at steep discounts on the secondary market. For example, several hedge funds, such as Golden Tree and Tilden Park, started buying up Puerto Rican debt in late 2017 and early 2018, taking advantage of low prices after Hurricane María. Before the storm, GoldenTree owned $587 million in public debt and increased its investments after María to $1.5 billion.[14] They now seek two, three, or almost four times a return on their investment. Creditors should not receive more than they paid. If they bought Puerto Rico's bonds for 14 cents on the dollar, they should not get more than 14 cents. The Commonwealth cannot afford to give vulture hedge funds a return on their investment if it means closing schools and slashing public health programs.

Despite these issues, the Fiscal Control Board, which has no interest in conducting a comprehensive audit, is now aggressively promoting restructuring agreements with bondholders, even after María. For instance, some of the restructuring agreements will force the island to pay double the principal on some loans. One example of this is the Puerto Rico Sales Tax Financing Corporation (known by its Spanish acronym, COFINA). COFINA is a public corporation that was created to issue government bonds and use other finance mechanisms in order to pay back and refinance Puerto Rico's debt. The particularity of COFINA is that it's main funding source is local sales tax. At present, a large portion of Puerto Rico's tax revenues goes directly to a private bank account where it is used to pay bondholders.

The COFINA agreement, tied to Puerto Rico's sales tax, proposes that close to $33 billion will be paid to bondholders through the maturity date of the debt.[15] But the principal—that is, the amount that the government actually received—was only about $17 billion.[16] That is not a restructuring of the debt. That is a gift to finance capitalists, particularly hedge funds, who bought parts of that debt for close to 10 cents on the dollar.

Additionally, some deals, such as the one proposed for the COFINA bondholders, make them the actual owners of Puerto Rico's sales tax receipts,[17] and will force us to keep on paying the tax for at least forty more years, without any of those funds going toward public services.[18]

Furthermore, the island is now due to receive a large influx of relief funds and external support to rebuild from the hurricane. This provides a real opportunity to bolster public investment and address the lack of efficient infrastructure that has contributed to the crisis. But as the Center for Economic and Policy Research has recently highlighted, "The Board's Fiscal Plan focuses on privatizing public institutions and putting funds aside for debt repayments, rather than using all available resources to help the island rebuild and recover."[19]

For all the foregoing reasons, we insist on a comprehensive citizen audit of Puerto Rico's debt. The Citizens' Commission—a natural successor of a previous government-created commission—is composed of seventeen experts and representatives in disciplines and areas impacted by Puerto Rico's public debt and the privatization and austerity measures advanced by its government and the Fiscal Control Board. The Citizens' Commission endeavors to examine the forensic, legal, constitutional, financial, and rendering components of debt issued by the Commonwealth of Puerto Rico, including its public corporations and retirement systems, since 1973. Thus, we are suing the government, seeking to obtain all public records needed to conduct such a project, including records that the government has voluntarily shared with the Fiscal Control Board and its bondholders.

Additionally, we are in the midst of a crowdfunding campaign, asking the public to help finance this much-needed audit.[20] In the end, our purpose is very simple: we wish to reclaim our right to learn and tell our fiscal, economic, and political history with regard to the accumulation of this odious debt, as well as to empower our people and social movements to change the power dynamics that keep our archipelago trapped by the twin evils of finance capitalism and imperialism.

The people before the debt, the people before austerity, a comprehensive audit before giving one more cent to finance capitalists!

1. I would like to thank Luis José Torres-Asencio for his comments and recommendations on this article.
2. Puerto Rico Oversight, Management, and Economic Sustainability Act (PROMESA), Public Law 114–87, 130 Stat. 549 (2016).
3. Omaya Sosa-Pascual, Ana Campoy, and Michael Weissenstein, "The Deaths of Hurricane María," Centro de Periodismo Investigativo, September 14, 2018, http://periodismoinvestigativo.com/2018/09/the-deaths-of-hurricane-Maria/.
4. Agustín Criollo Oquero, "100K Signatures Collected in Favor of Auditing Puerto Rico's Debt," *Caribbean Business*, February 21, 2017, https://caribbeanbusiness.com/100k-signatures-collected-for-the-audit-of-puerto-ricos-debt/. See also Carla M. Pérez Meléndez, "Encuesta revela apoyo a la auditoría de la deuda," *Diálogo UPR*, April 4, 2017, http://dialogoupr.com/encuesta-revela-apoyo-la-auditoria-de-la-deuda/. The signatures in favor of a comprehensive audit of Puerto Rico's public debt have since surpassed the 150,000 mark.
5. 80grados, "La auditoría va, pero ciudadana," April 20, 2017, https://www.80grados.net/la-auditoria-va-pero-ciudadana/.
6. Armando J. S. Pintado, "Pausa para la Auditoría," *80grados*, March 24, 2017, https://www.80grados.net/pausa-para-la-auditoria/.
7. For a discussion on the importance of citizen audits as strategies for public education, participation, and mobilization, see María Lucia Fattorelli, *Auditoría ciudadana de la deuda pública: Experiencias y métodos* (Brasilia: Inove Editora, 2013).
8. Centro de Periodismo Investigativo, "Deuda pública, política fiscal y pobreza en Puerto Rico," April 4, 2016, http://periodismoinvestigativo.com/wp-content/uploads/2016/04/FINAL-Informe-Audiencia-Pública-PR-4-DE-ABRIL-2016.pdf (presented by civil society organizations to the Inter American Human Rights Commission); José M. Atiles-Osoria, *Apuntes para abandonar el derecho: Estado de excepción colonial en Puerto Rico* (San Juan: Editora Educación Emergente, 2016); Naomi Klein, *The Battle for Paradise: Puerto Rico Takes on the Disaster Capitalists* (Chicago: Haymarket Books, 2018); Ricardo Cortés Chico, "La austeridad como receta," *El Nuevo Día*, April 22, 2017, https://www.elnuevodia.com/noticias/locales/nota/laausteridadcomoreceta-2313854/; Roberto Pagán, "La austeridad y las reformas antiobreras no son la respuesta," *80grados*, December 8, 2017, https://http://www.80grados.net/la-austeridad-y-las-reformas-antiobreras-no-son-la-respuesta/.
9. *El Nuevo Día*, "Pautada la primera reunión de la Comisión de Auditoría de Crédito Público," January 7, 2016, https://www.elnuevodia.com/noticias/politica/nota/pautadalaprimerareuniondelacomisiondeauditoriadecreditopublico-2147966/.
10. Puerto Rico Law No. 22–2017. See also Leysa Caro González, "Aprueban a viva voz derogar comisión que examinaría la deuda pública," *El Nuevo Día*,

April 18, 2017, https://www.elnuevodia.com/noticias/locales/nota /caldeadoslosanimosenelcapitolio-2312117/.

11. See Pagán Rodríguez et al. v. Rosselló-Nevares, SJ2017CV00037 (Judgment of April 6, 2017) (Puerto Rico Court of First Instance, Superior Court of San Juan) (ordering the restitution of the commission's citizen representatives, previously dismissed by Governor Rosselló-Nevares).

12. Comisión para la Auditoría Integral del Crédito Público (CAICP), *Pre-audit Survey Report* (San Juan: CAICP, 2016), http://periodismoinvestigativo.com /wp-content/uploads/2016/06/Informefinal.pdf

13. Saqib Bhatti and Carrie Sloan, "Puerto Rico's Payday Loans," *ReFund America Project*, June 30, 2016, https://bibliotecavirtualpr.files.wordpress .com/2017/04/2016-2017-refund-america-debt-pr-english.pdf.

14. Abner Dennis and Kevin Connor, "The COFINA Agreement, Part 2: Profits for the Few," *Eyes on the Ties*, November 20, 2018, https://news .littlesis.org/2018/11/20/the-cofina-agreement-part-2-profits-for-the-few/. The updated figures are reflected by the creditor's own filings in the ongoing judicial proceeding under PROMESA; see *In re*: The Financial Oversight and Management Board for Puerto Rico, No. 17 BK 3283-LTS.

15. Joanisabel González, "Bajo la lupa de Swain el plan de COFINA," *El Nuevo Día*, November 18, 2018, https://www.elnuevodia.com/noticias/locales /nota/bajolalupadeswainelplandecofina-2460418/.

16. González, "Bajo la lupa de Swain."

17. Dennis and Connor, "The COFINA Agreement."

18. Dennis and Connor, "The COFINA Agreement."

19. Lara Merling and Jake Johnston, *Puerto Rico's New Fiscal Plan: Certain Pain, Uncertain Gain* (Washington, DC: Center for Economic and Policy Research, 2018), http://cepr.net/images/stories/reports/puerto-rico-fiscal -plan-2018–06.pdf.

20. To follow our crowdfunding campaign, visit http://www.auditnow.org/.

RHIZOMATIC

Ana Portnoy Brimmer

*"By Porto Rican law the entire beach of the island is govern-
ment property, for sixty feet back of the water's edge. As a consequence, what would in our own land be the choicest residential
section is everywhere covered with squatters, who pay no rent,
and patch their miserable little shelters together out of tin
cans, old boxes, bits of driftwood, and yagua or palm-leaves,
the interior walls covered, if at all, with picked-up labels and
illustrated newspapers."*
—Harry A. Franck, *Roaming through the West Indies*, 1920

Mangroves now lie on the side of the road.
Cut to charred bones and left to rot

on the dirt path leading to the bay,
where people from the capital,

or from the continent, have been
building second or third homes.

As if the waterfront wasn't home
already. Long ago, times only the salt

can claim to have clung to, these edges
of land were shunned, deemed disease-ridden,

too poor to possibly inhabit, too dark
a stretch of marginal acreage.

A blue expanse was anything but useful,
laden with reef and memory.

But our story begins here. With water.
The one constant, always calling.

They have many names for us, you see.
Squatters, informal dwellers, deedless

in square-footage. But these coasts have long
known our laborious footfall, sand swallowed

our prints and those of our ancestors,
soil seen the harvest and burn of cane

and our children. Like mangroves,
we are of an intertidal existence—

led by the comings and goings of the blue,
of brine and resolute rooting.

Our fingers are tattooed with clasp
of land crab, prying of clams and oysters,

nibble of fish and estuary waters,
sharp imprecision of machete on coconut.

This land is no one's. This land is of itself—
the lick and lip of wave's edge.

But it has taken us in. Made a home
of us in return. So, we too taste the pierce,

the bleeding fence—acid whispers
of *private* and *profit*. They want us gone.

Offer to repair our homes from the storm,
if only we move inland—recede, swell-like,

from the shore. Let another hurricane come.
We stand rhizomatic upon this saline earth.

PART V
Transforming Puerto Rico

LOOKING FOR A WAY FORWARD IN THE PAST

Lessons from the Puerto Rican Nationalist Party

Mónica Jiménez

In June 2016, the US Supreme Court ruled in two important cases on the disputed sovereignty of Puerto Rico. These decisions, though significant, were quickly overshadowed by Congress's passage of the Puerto Rico Oversight, Management, and Economic Stability Act (the much-reviled PROMESA), which sought to address the island's billions of dollars of debt. Collectively, these actions reasserted the United States' plenary, or complete, power over the island and reinscribed the face of US colonialism in Puerto Rico. In the wake of the staggering debt and economic crises the island faces, I began to ask myself what if anything we might find in the past, and specifically in the ideas of the Partido Nacionalista de Puerto Rico (Nationalist Party) and its leader, Pedro Albizu Campos, that might help us understand this moment and guide us into the future.

Then, as the world watched in horror and grief, what was already a seemingly insurmountable problem of debt and colonialism became a catastrophe. Hurricane María hit Puerto Rico in September 2017 and laid bare for all to see the vulnerabilities and inequities of Puerto Rican life. The hurricane exacerbated the many daily difficulties islanders already faced in merely trying to live. Puerto Rico, nearly invisible to most of the world before the hurricane, became hyper-visible in the wake of disaster, and so too did colonialism in its twenty-first-century iteration.

In the face of so many catastrophes and when the ideas of sovereignty and democracy no longer hold any real value for the island, where can Puerto Ricans find a way forward? Where can we find guidance or other ways of thinking about potential futures? For an island that has been colonized for over five hundred years, this is a very difficult question indeed. In light of the island's current obstacles, the need to think creatively seems even more urgent, as does turning to the island's history and thinking through what we might glean about the way forward. What guidance might we find in the old ideas—those that, at the time, were tossed aside as impossible, ill conceived, retrograde, or dangerous?

The notions of sovereignty and autonomy have a rather slippery history for the island. Before the arrival of the United States, islanders lobbied for greater autonomy from Spain, which they received mere months before the US occupation of 1898. In point of fact, the island has never truly had sovereignty, having always been held as a colony. In the mid-twentieth century, Albizu Campos called vociferously for independence for Puerto Rico, arguing that anything less than full sovereignty would be tantamount to colonialism. That period was particularly fraught, as the Great Depression and the downturn of global markets had made Puerto Rico and its export-based economy particularly vulnerable. In the midst of increasing desperation on the island, Puerto Ricans began to demand a change to their relationship with the United States and an end to exploitative labor practices.

In 1952 the creation of the Estado Libre Asociado (ELA, or the Commonwealth of Puerto Rico) seemed to respond to islanders' calls for change. The ELA created a local government to oversee island affairs and allowed islanders to elect their own local officials. It operationalized the notion of local autonomy in close association with the United States. Even after the creation of the ELA, however, the question of whether the island had acquired something like sovereignty loomed large. Over the past sixty years this has been a point of debate in local political circles, but in June 2016 the US Supreme Court gave islanders a definitive answer with its decisions in *Puerto Rico v. Sanchez Valle* and *Puerto Rico v. Franklin California Tax-Free Trust, et al.*[1]

These two cases asked the court to rule on the limits of the island's ability to govern itself. *Sanchez Valle* asked the court to decide whether Puerto Rico was a "separate sovereign" for purposes of the Fifth Amendment's double-jeopardy clause, which allows for two prosecutions to stem from the same series of events if the prosecutions are pursued by "separate sovereigns." In that case the court found that the power to prosecute crimes stems from the island's constitution, as is true for all fifty states. But unlike the states, whose constitutions stem from the powers granted to each by the people who reside in them, Puerto Rico's constitution came about from a delegation of congressional power. As a result, the island's power to prosecute criminals stems from Congress, and where there is a dispute over whose power reigns supreme, that of Congress wins out.

Puerto Rico v. Franklin California Tax-Free Trust, et al. asked the court to decide whether the island can pass its own bankruptcy law to address its overwhelming debt. There, the island argued that since it is not a state, it is not barred from passing its own bankruptcy laws. The court disagreed and found that it was not up to Puerto Rico to decide when its indebted governmental entities can pursue bankruptcy, as is the case with the fifty states. Instead, it is up to Congress to decide on that matter. Together, these two decisions reaffirmed Congress's plenary, or complete, power over Puerto Rico, which was established by the 1901 Supreme Court decision *Downes v. Bidwell.* There, the court pointed to Article IV, Section 3 of the US Constitution, the Territory or Property Clause, to find that Congress had plenary power over the island and as such Congress could choose if and when to extend the provisions of the Constitution to Puerto Rico. The court also created the unique designation of "unincorporated territory" for Puerto Rico, which meant that the island was outside the established legal order and could be maintained as such until such time as Congress chose to change that status.

These decisions, in tandem with the passage of PROMESA and its implementation of an unelected Fiscal Control Board to oversee the island's financial affairs, effectively spelled the death of the ELA and the island's limited self-governance. They also signaled a return to the sort of colonial relationship that had existed before its creation. The events of 2016 would no doubt have been unsurprising for Albizu Campos, who in many ways predicted much of what is playing out in the island's political and economic sphere today. His warnings against embracing the ELA and what ELA would actually mean for the island were quite prescient.

When ELA was created, he urged Puerto Ricans to reject it as just another form of colonialism but with a shinier veneer. In 1950, when islanders were asked to vote to approve the legislation that would enact the island's constitution, Albizu Campos stated, "In Puerto Rico only the US has jurisdiction . . . That Law [Law 600, which enacted the ELA] can be annulled by the US Congress."[2] He went on to argue that the island's purported constitution would be subsumed by Congress's power. As an attorney, Albizu Campos understood well the limitations that Congress's plenary power placed on the island, and he also understood that creating a constitution for Puerto Rico did nothing to abrogate that power. For him the constitution was merely smoke and mirrors meant to placate Puerto Ricans without actually changing the colonial dynamics of the island's relationship with the United States.

Perhaps one of the points Albizu Campos made most vehemently was the impossibility of democracy for Puerto Rico within a US context. Indeed, the logic of the ELA was that there could be local democracy despite federal plenary power. In 1933 he wrote, "In Puerto Rico there is no legislative power."[3] In other words, although the island had institutions that mirrored those of the federal and US state governments—a Congress on the island, which was elected by the people and was charged with representing their interests—they were nonetheless powerless before the veto power of the US-appointed governor, Congress's representative on the island. He argued that while US plenary power existed, the people of the island could never have true democracy. Later, supporters of the ELA and of Luis Muñoz Marín, the island's first democratically elected governor, would point to this election and the island's adoption of its own constitution to refute Albizu Campos's arguments.

The Supreme Court's 2016 decisions with respect to the origins of the island's power to self-govern effectively curtailed Puerto Rico's local democracy and reaffirmed plenary power—it also proved Albizu Campos's right all those years later. In 2016 the court held that the island's Constitution and its ability to self-govern are merely delegations of congressional power and are thus limited. In other words, what Congress gives, Congress can take away; when in doubt, Congress's power supersedes that of the island government and the people of Puerto Rico. Although the ELA allows for local governance, how effective is local democracy if ultimately Congress can veto island officials and assert its will in even the most local facets of island life?

In March 1933, while Puerto Rico was still in the process of recovering from the catastrophic Hurricane San Ciprian, which struck the island in 1932, Albizu Campos wrote to Rafael Martínez Nadal, the president of the Puerto Rico Senate, urging him to adopt measures to stop the island's payments on its debts to US-held banks.[4] Then as now, the island was in the midst of a major financial calamity brought on by the Great Depression and two major hurricanes. Albizu Campos's March 7 letter came in response to President Roosevelt's implementation of a national bank holiday and passage of the Emergency Banking Act of 1933 (EBA), which were intended to protect banks from the effects of mass withdrawals as people panicked when the economy hit bottom in the winter of 1932–33. The bank holiday and other protections implemented by the federal government through the EBA were intolerable to Albizu Campos, who decried the fact that those same banks had not extended relief from mortgage or other debt payments to devastated islanders after the back-to-back hurricanes.[5]

Albizu Campos argued that the US financial system could not help but take advantage of the vulnerabilities brought on and worsened by the hurricanes. He warned that Puerto Ricans would be naive to look to the banks to help the island recover from the natural disasters. For him the post–San Ciprian moment and the possibility of government intervention to help the banks posed a golden opportunity to consolidate power over the island's finances in the hands of US corporate interests. Ultimately, Albizu Campos's arguments about the predatory nature of the US banking system seemed to anticipate exactly the financial debacle the island faces today. His warnings against a system that protected big banks but not devastated citizens also resonate in Puerto Rico's current moment. While the world has changed much since 1933, Albizu Campos's fears about the consolidation of the island's finances in the hands of US bankers and the loss of what he called "the island's natural riches"— but which we might understand as its public sector, its public lands, its *res publica*, or public thing—anticipated the slow-moving train of the island's slide into debt financing and its current condition.

While there might not have been much radical about what Albizu Campos was saying about the island's lack of legislative power or his warnings about predatory bankers or even his declaiming the possibility of local democracy in association with the United States, nevertheless his warnings fell on deaf ears. Perhaps what Albizu Campos saw when he considered the island's relationship with the United States in the early twentieth century was actually the future—our present. A future that seemed so unimaginable to his contemporaries that they simply dismissed his warnings as the ravings of an extremist, easy to ignore and dismiss.

Albizu Campos was also adamant that the solutions to the island's problem of colonialism would not be solved by the United States. In fact, he declared that "we will never arrive at being a free people if we keep *asking* the US what should become of us."[6] In this I agree with him; over the past 120 years, appeals to Congress have been ineffective in resolving some of the island's most pressing problems. In 2016, prominent Puerto Ricans of all stripes, including politicians, academics, and celebrities, lobbied Congress to do something, anything, to address the island's debt crisis. The resulting legislation, PROMESA, was a disappointing answer to those pleas. And of course, in September 2017, when Hurricane María hit, as all the world watched, desperate islanders pleaded for aid from the United States and were met with neglect, incompetence, and paper towels. Again, the US response fell short and proved once again that when islanders seek the aid of the US government, what they can expect in response is more colonialism.

It has been eighty-six years since Albizu Campos wrote to Martínez Nadal calling on him to halt Puerto Rico's debt payments and indicting the predatory nature of the US banking system. Eighty-six years later, Puerto Ricans are regularly taking to the streets, to social media, and to La Fortaleza, the Puerto Rico governor's mansion, to demand the very things that Albizu Campos called for in the 1930s. Albizu Campos warned us not to be lulled into believing that the US could offer the island something other than colonialism. Despite the intervening years, the apparent changes in Puerto Rico's political status, and in the face of the devastation wrought by Hurricanes Irma and María, Albizu Campos's warning rings truer today than it did even in the late 1940s, when the island was abuzz with the possibilities of Estado Libre Asociado.

1. *Puerto Rico v. Sanchez Valle*, 36 S.Ct. 1863 (2016); *Puerto Rico v. Franklin California Tax-Free Trust, et al.* 136 S. Ct. 1938 (2016).
2. Cited in Marisa Rosado, *Pedro Albizu Campos: Las llamas de la aurora, acercamiento a su biografía* (San Juan: Ediciones Puerto, 2008), 350.
3. "Letter to Rafael Nadal of March 7, 1933," in *Pedro Albizu Campos: Obras escogidas, 1923–1936*, vol. 1, ed. J. Benjamín Torres (San Juan: Editorial Jelofe, 1975), 241.
4. "Letter to Rafael Nadal of March 7, 1933," in *Pedro Albizu Campos*, 241.
5. "Letter to Rafael Nadal of March 7, 1933," in *Pedro Albizu Campos*, 241.
6. El Mundo, "El estado federal para PR no es acceptable porque destruira nuestra personalidad colectiva," January 31, 1923, in Torres, *Pedro Albizu Campos*, 14 (emphasis in original).

PSYCHOANALYSIS AS A POLITICAL ACT AFTER MARÍA

Patricia Noboa Ortega

When I was invited to participate in the panel "Transforming María," I asked myself the following questions: What should we transform? Can we transform our collective and individual pain? If so, how? Based on these questions, I will describe, as a psychologist, the type of support I offered in communities in Puerto Rico after María's impact, and as a researcher, I will present ethnographic research that I conducted in one community and share some of the results.[1]

A month after María struck the island, I, like many Puerto Ricans, was dealing with a sense of loss, impotence, and anger. These feelings arose every time I heard on the radio or read in the news about the local and federal governments' poor response to the disaster. To transform those emotions, I decided to participate in an initiative by the Puerto Rican College of Physicians and Surgeons, which organized health brigades to visit affected communities and offer health services. The brigades were composed of family doctors, pediatricians, psychologists, nurses, and medical students, among others.

PSYCHOANALYSIS AS A POLITICAL ACT

I went to the communities of Humacao, Canóvanas, and Utuado. I am a psychologist with a background in psychoanalysis, so as a member of the health brigades, I tried to provide a space in which residents could talk about their experiences and thus give them the opportunity to calm their anguish and discomfort. I tried to sustain a space for them to talk about their distress, suffering, and sense of abandonment.

As psychoanalysis opens a space for human beings to calm their anguish through words, it also makes it possible for another person to respectfully listen without giving a diagnosis or prescribing anything. This is a political task, since psychoanalysis views the subject in its singularity and acknowledges that what provokes suffering in each subject is unique. For example, during the hurricane, two human beings could have lost the same object (car, bed, house, photographs, etc.), but the response to each loss is different. Perhaps one cried inconsolably for a long time, without sleeping or eating properly, and each time the thought of the object came to mind, one experienced an acute pain in the chest, a state of anguish. To the field of psychoanalysis, the cause of this suffering does not lie solely in the loss of the object; the object in itself does not carry meaning, which results from the representations that the subject makes of the lost object.

In my case, I remember the noise of the winds coming through my windows during the storm, blowing with all their strength against my sliding doors. Recklessly, I confronted the winds, pushing the sliding doors back in place. I knew I was risking my life, but I was also trying to protect everything I had worked so hard for throughout the years: my home, which was much more than a sofa, dining table, or kitchen. That space represented the promises I made to myself in my childhood. I wanted to become an independent woman who could provide for herself, someone who was not going to live in poverty anymore. Those signifiers, or representations, gave me the strength to defend and protect my home while risking my life. For a psychoanalyst to pay attention to that kind of psychic reality is a political act, as long as the subject's grief is addressed in the therapeutic process.

Much of human suffering is caused by the excessive exigencies imposed on people when they identify with cultural ideals, such as that one must be happy or "stand out" in one's discipline. Psychoanalytic practice allows the analyst to disidentify from such cultural exigencies while helping the subject assume a different, more critical position toward them.[2] As Willy Apollon, a Haitian philosopher and psychoanalyst, points out in his text *What Is Psychoanalysis?*, "It is the possibility of calming the anguish through a word that is articulated and understood. It is the possibility of treating the symptoms, the unexplained sense of pain. It is the possibility of discovering what directs our life and that which we ignore. Psychoanalysis is a task drawn from the unconscious."[3]

An Argentine psychoanalyst, Juan David Nasio, in his book, *Yes, Psychoanalysis Cures* (2017), asserts, "To listen to the speaker is to concentrate actively on what he says, to try to go beyond the words he utters and, above all, more than anything else, to feel his conscious emotion in us and if possible, his painful and unconscious emotion." Let me share one experience that illustrates my act of listening.

The brigade was offering its services at a church in Punta Santiago, a poor community in Humacao, on the east coast of the island. There, I met a woman who was about sixty-five years old. She was married and had an adult daughter. She and her husband were fishers and had a fish shop. At the church, she was handing out coffee to those who were waiting to be seen by the doctors. I greeted her warmly and asked her, "Elisa, how are you? How is this leader?" (she was one of the leaders who helped organize the activity.) She looked at me with surprise and asked, "How do you know my name?" I told her it was written on her name tag. She looked at her chest, laughed, and talked to me. She told me how difficult the hurricane had been for her. She named all the things she had lost. "I lost my clothes, my shoes, my washing machine, my dryer, my car, my fish for sale, and the refrigerator where I put the fish," she said.

Right away, she added, "Those are material things. They do not matter. At least we are alive." I answered, "Is that so?" This question seemed to trigger something in her, and she started to cry. Immediately, she said, "I cannot cry. Here, they need me. I must be strong. I cannot become lost. I cannot lose my mind." I replied, "Tell me more about what you lost." Elisa started telling me about the bed she had lost and how losing it had forced her to sleep at the parish church, and for her doing this was like "invading" that space. She was "invading the preacher's house," and she did not want to feel this way. Elisa "did not want to bother" anyone. She told me all this, crying and full of grief.

I want to highlight significant aspects of her narrative and how I listened to her pain. First, my question "Is that so?" opened a space for her to elaborate in her own words the pain and suffering she was going through, the loss of the bed, how that loss led her to sleep at the parish church, and how she signified that as an "invasion." María has definitely been a traumatic event for everyone in Puerto Rico. Each of us, however, experienced and gave meaning to the event and our losses in a unique way. In the psychoanalytic approach, we listen to the present trauma (Hurricane María), but we also listen to past traumas. These are unique experiences that the subject has never named or spoken of, but now María has triggered them.

Second, my question "Is that so?" responded to the discourses of resilience that appeared after María, such as "Puerto Rico se levanta" (Puerto Rico rises up) or "Vamos pa'lante" (Onward). These discourses emphasize individual responsibility, but they do not necessarily reflect the government's responsibility toward us. When we identify with these discourses, they become exigencies in our mourning process. I want to underline two exigencies that Elisa had toward herself:

1. *She had to be a strong woman who resists crying.* For her, crying was a sign of weakness, and it could make her lose her mind. To understand her worry over "losing her mind," it is important to mention that she had received prior psychiatric treatment as a consequence of losing her job. For the past thirteen years, because of the economic recession in Puerto Rico, we have lost 272,000 jobs, and serious mental illness afflicts 7.3 percent of adults (eighteen to sixty-four years old). In other words, more than 165,000 adults need mental health services.[4][5]

2. *She had to be attentive.* She had to respond to the needs of her family, her community. But she seemed to refuse to acknowledge her own needs, and when they were addressed, she experienced anguish.

This case illustrates what we have been living through in Puerto Rico before María, such as losing jobs because of the economic crisis. Community leaders, especially women, have been taking care of family and community needs, dealing alone with their pain and with the limitations of psychiatric treatment, much of which relies on psychotropic medication.[6] This medication treats the biochemistry of the organism, operating exclusively in the biological field. But it does not treat the causes of depression, such as the excessive exigencies imposed on the subject to comply with a cultural ideal.[7] In psychoanalysis the illness in the subject is not conceptualized as a medical object; the subject suffers because he or she is in conflict with him- or herself or with others. Precisely through analysis, those interior or exterior conflicts begin to disappear. From a psychoanalytic point of view, a person is "cured" when they can love themselves as they are, when they have become more tolerant of themselves and their surroundings and, therefore, also more tolerant of others.[8] This approach to mental health is political as long as the clinical work disengages from the ideals promoted by capitalist culture.

ETHNOGRAPHY AS A POLITICAL ACT

My ethnographic work was conducted in Valle Hill, a sector from the San Isidro community in Canóvanas. San Isidro is located near the flood plain of Río Grande de Loíza, the most abundant river on the island. Valle Hill is located in a flood zone.

According to the 2010 Census, 6,288 residents live in San Isidro; 53 percent are women, 53 percent are unemployed, 46 percent are employed (in service and office occupations, as well as in retail, manufacturing, construction, and waste management), 67 percent obtained a high school diploma; 95 percent are Puerto Ricans, 2.1 percent are Dominicans, and 59 percent live below the poverty level.[9] Residents do not have a sewer or a potable-water system; they do not have a community center, Head Start program, library, or sports or recreational facilities.[10] About thirty-five hundred residents live in Valle Hill, an area of San Isidro. Many are immigrants from the Dominican Republic.[11] Some have been living there for the past twenty-five years. Many came to Valle Hill because a former mayor, José "Chemo" Soto, made available plots of land to construct houses in hopes of "a new beginning." Throughout the years, residents, with the support of the municipal government, have been illegally filling the wetlands to construct homes. Because of those violations, the

Environmental Protection Agency penalized the municipal government with a $128,000 fine.

The poor-quality materials used, like tin roofs and deteriorated wood, and the informal construction practices coupled with living over a wetland, have created conditions of vulnerability toward a weather event like María. I came to Valle Hill in October 2017, a month after María. The level of destruction there had such an impact on me that I decided to come back. About 41 percent (900) of the houses were completely destroyed.[12] In Puerto Rico, two hundred and fifty thousand homes were severely damaged.[13]

In November, I met with the Valle Hill Community Board members. Jannette Lozada, the president of the board, started the meeting talking about the loss of her mother, who died during Hurricane Irma. Alberto, another board member, told me about the death of his neighbor (who died of leptospirosis), and how he almost drowned during María (he stayed home during the hurricane). These expressions refer to death—the loss of a mother and a neighbor, and the possibility of dying oneself, thus illustrating the collective trauma that was experienced in Puerto Rico after María. Some three thousand people died, according to estimates, mostly because the health system collapsed. About three-quarters of hospitals were not operating, treatment clinics were closed, and over one thousand medical offices and pharmacies were not providing services. All in all, this was because the island's general infrastructure collapsed—there was no electricity, running water, or telecommunications. Three months after María, many people died because of accidents, heart attacks, diabetes, sepsis (because of high temperatures in the hospitals and unsanitary conditions), suicides, respiratory conditions, and high blood pressure. Twenty-six people were certified as having died from leptospirosis; most of these deaths occurred from September 2017 to March 2018.[14]

In this immediate context of destruction, and in the long-term context of poverty, exclusion, and discrimination in Valle Hill San Isidro and throughout Puerto Rico, I wanted to document and highlight how the Valle Hill residents recovered from Hurricane María through their collective actions, on their own terms. I listened to their narratives and reflected on the experiences they shared with me and how and why they did so. I was also open to their interests and needs. The health brigades' response, including providing services and distributing donated supplies, had to be fast because the priority was to meet the residents' basic needs, not carry out research. Nonetheless, I was able to carry out my research, focusing on three main questions: (1) What challenges did people face in their recovery process? (2) How did they cope with those challenges? (3) What psychosocial effects were they dealing with? I relied on participant observation, field notes, ethnographic interviews (unstructured and structured), and ethnographic narratives.

Acknowledging that basic needs must be met to begin the recovery process, we started identifying the residents' needs and brought donations (water filters, clothes, medications, furniture, and food). We also brought a lawyer to help them appeal FEMA decisions. We coordinated health brigades for the bedridden and the elderly, like Abuelo, who had not received medical attention for the past seven years. We held an art activity for children. We wanted them to draw their experiences of Hurricane María. Their mothers told us they were very anxious. Because of the sustained and respectful support we gave them, we built a trustful relationship with the residents of Valle Hill.

Before María hit Puerto Rico, the municipal government opened shelters, but because many of them lacked food and water and had poor hygiene, residents decided to stay in their community. They trusted their immediate neighbors, not the municipal government. Those residents who stayed in their homes or neighbors' houses encountered over fifteen feet of water (which was contaminated from the wetland and with sewage from flooded septic tanks). Residents were rescued by boats and improvised rafts, such as refrigerators. Some neighbors even had to swim from their houses to rescue these residents. Because of the flooding, residents lost all they had. As Lulú, a Dominican woman, said, "I cannot forget that image when I came by and saw all the houses destroyed, but the big thing for me was when I arrived here [referring to her house]. There was nothing. Everything was gone." In this context of loss and destruction, facing years of social exclusion as a result of racism, discrimination, and stigmatization, the Valle Hill residents started their recovery process.

Several residents have been discriminated against because they are poor, Black, and do not speak English. Many residents commented that FEMA inspectors spoke little or no Spanish. Others reported that they were mistreated by the inspectors: they were told to be silent, the inspectors joked about the situation, and they received FEMA correspondence in English, even though the residents had requested them in Spanish. Other residents did not receive guidance on the appeal process. All this constituted an obstacle for residents with limited English.

The case of Lulú illustrates the federal authorities' discrimination against the community. Lulú has lived in Puerto Rico for several years, works in a school, and is a cosmetology teacher. She lost her home, which also served as her studio. She made an album to document her losses and facilitate FEMA's inspection. The album included a photograph in which the mannequin heads she uses in her cosmetology courses appeared. When a FEMA inspector, who knew little Spanish, came to her house, she sarcastically asked Lulú, "Are those heads FEMA inspectors' heads?" In other words, she was suggesting that Lulú was practicing witchcraft against FEMA inspectors. When Lulú saw my face of indignation, she told me, "Hey, they can cut the branch but not the trunk" (Nos pueden cortar las ramas, pero no el tronco), suggesting that this insensitive and disrespectful comment did not break her. Valle Hill is not the only community that suffers discrimination; other communities have also been discriminated against in Puerto Rico.[15] When you feel discriminated against, when someone treats you disrespectfully, as though you are dishonest or a less intelligent person, this negatively affects your psychological health, exacerbating feelings of sadness, abandonment, anger, and loneliness.[16]

Pages from Lulú's album. Photo by author.

Valle Hill's recovery was undertaken by residents whose houses were damaged, along with neighbors, family members, nonprofit organizations (Techo, Red Cross, Catholic and evangelical churches Amma—Embracing the World—Eaton), pri-

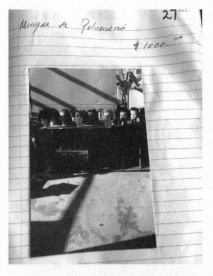

vate and financial entities (Banco Santander), cooperatives (Cooperativa Ecológica de Permacultura para un Puerto Rico Autosuficiente y Cooperativa de Ahorro y Crédito Roosevelt Roads, and citizen groups (like Brigada de Todxs). Some of them brought donations, installed tarps, or helped rebuild houses. Some residents shared meals they prepared and built a community kitchen; other residents shared donations they received with their close neighbors. Some offered their own homes as shelters (for months) or helped clean their neighbors' houses. Some residents cleaned the streets, which were full of debris, or cleaned up the wetland.[17]

Brigade members assess a damaged home. Photo by author.

As in many communities in Puerto Rico, Valle Hill community members raised money, procured electrical equipment, and reconnected to the electrical grid. These collective actions of survival, however, did not prevent the deterioration of the residents' health. Diabetes, cancer, hypertension, gastrointestinal diseases, asthma, allergies, and skin diseases led some residents, like Cano, to die. Cano had respiratory problems and sleep apnea. Both conditions required medical equipment, which he could not use after the storm because there was no electricity. Cano's case illustrates how preexisting conditions, without equipment, electricity, and access to medical treatment, caused deaths. When I heard about his death, I visited Ana, Cano's wife. When she saw me, she hugged me strongly and told me crying, "I could not save him. I lost him."

Cano before his death. Photo by author.

The residents attended to supplying their immediate needs, such as water, shelter, food, clothes, and medication, while expressing their suffering to others. It seems to me that narrating that suffering served as a political strategy to attain a necessary good in the recovery process. For example, on many occasions, residents would cry because of what they had lost. One resident said, "This Christmas will be the worst because I will not have my children with me." This resident's children had migrated to the United States, and the street where they lived was now empty. As a result, this resident not only lost family but also neighbors who left their houses; it was a desolate scene indeed. Another resident commented, "I lost everything. The only thing I have left is what I have on [the clothes she was wearing] and these sandals." This resident spent the hurricane at the hospital because she had no one who could pick her up; she spent several days in anguish, feeling helpless, and not knowing if her house had survived the storm. A community leader would often say, "My community is being emptied," referring to all the residents who had migrated to the United States or returned to their families in the Dominican Republic. About one hundred and sixty thousand Puerto Ricans migrated to the United States after the storm, according the Center for Puerto Rican Studies. This exodus represents one of the most significant relocations of Puerto Ricans to the US mainland in history.[18]

Ana María Fernández, an Argentine psychoanalyst, once said, "We cannot psychologize the social."[19] In other words, we should recognize how culture, economics, and politics cause suffering among our citizens. If we want to transform our collective pain, we must work toward changing those circumstances and understanding the subjective causes, the ones that are based on the singular experiences each of us have had with Hurricane María in Puerto Rico. Therefore, it is of utmost importance to continue opening spaces to listen to the residents' pain. That is why we developed a Legal and Psychological Clinic (LPC)[20] to address the legal problems residents have faced by accessing the economic assistance they are entitled to, and also sustaining a space for them to talk about the pain, sadness, and adversities they have struggled with in the aftermath of disaster.

The legal component of the LPC's work is to help participants perceive themselves as (1) knowledgeable of their rights and the steps they have to follow for their recovery or displacement process, (2) more prepared to identify their needs, and (3) able to use the legal tools introduced to them in the clinic to identify solutions to the problems they have faced during the recovery process. In community meetings and orientations, we discussed residents' shared challenges in dealing with the municipal, local, and federal governments. For example, FEMA is asking Valle Hill residents for a certification issued by the Department of Housing to prove that they live there. In the past, this certification was issued by the municipal government, and residents had to apply every six months. Now, because more obstacles are interrupting access to economic assistance, the process is more difficult for them. We also conducted individual legal counseling by assisting residents with their particular situations.

The LPC's psychological component aims to help participants perceive themselves as more competent to (1) deal with the adversities, emotions (anger and sadness), and psychological process (e.g., mourning) they have experienced, and (2) to manage the difficulties, both emotional and psychological, during their recovery process. Both the legal and psychological components are addressed through community group discussions. We want to create a common space in which people can talk to each other about their suffering. Talking with one's neighbors creates proximity and trust, and it enhances one's social support network. We also offer individual clinical sessions (up to five sessions per participant), in which people can elaborate through words the pain and sadness they are dealing with. This individual work helps them gain insight into their emotional and psychological dimensions, which are in a way unknown to them.

1. This chapter is based on a paper presented at the New Brunswick Theological Seminary at a conference titled "Aftershocks of Disaster: Puerto Rico a Year after María." at a panel titled "Transforming María," at Rutgers University, New Brunswick, New Jersey. First published as "Psychoanalysis and Research for Communities," *Cruces*, November 15, 2018. Research was supported by the National Institute of General Medical Sciences of the National Institutes of Health under Award Number R25GM121270. The content is solely the responsibility of the author and does not necessarily represent the official views of the National Institutes of Health.

2. For an in-depth analysis, see Jorge Alemán, *Lacan y el capitalismo, Introducción a la Soledad: Común* (Granada: Universidad de Granada, 2018).

3. Willy Apollon, *¿Qué es el psicoanálisis?* Escuela Freudiana de Quebec. For an in-depth analysis, see Willy Apollon, Danielle Bergeron, Lucie Cantin, "Tratar la psicosis," in *Tratar la psicosis* (Polemos, Buenos Aires, 1997). Also, Willy Apollon, Danielle Bergeron, Lucie Cantin. *After Lacan: Clinical Practice and the Subject of the Unconscious* (New York: State University of New York, 2002).

4. Jose Caraballo, "El efecto exacerbante del huracán María sobre la quebrada economía de Puerto Rico," Oral presentation in Symposium of Huracán · María at the University of Puerto Rico Cayey, 2019.

5. Glorisa Canino et al., "Need Assessment Study of Mental Health and Substance Use Disorder and Service Disorder and Utilization Among Adult Population of Puerto Rico," Behavioal Sciences Research Institute Final Report, December 15, 2016.

6. For an in-depth analysis, see María Dolores-Fernós, Marilucy González-Báez, Yanira Reyes-Gil, and Esther Vicente, "Voces de mujeres: estrategias de supervivencia y de fortalecimiento mutuo tras el paso de los huracanes Irma y María," Inter-Mujeres, 2018.

7. Elisabeth Roudinesco, *¿Por qué el psicoanálisis?* (Buenos Aires, Argentina: Paidós, 2007).

8. Juan Nasio, *Sí, el psicoanálisis cura* (Buenos Aires, Argentina: Paidos, 2017).

9. Características económicas seleccionadas comunidad San Isidro; Encuesta sobre Comunidad de Puerto Rico del 2012–16. Estimado de cinco años. Tabla DP03.

10. Oficina de Comunidades Especiales, *Perfil Socioeconómico de la Comunidad de Valle Hill*, 2003.

11. Jannette Lozada, Valle Hill community leader.

12. Jannette Lozada, Valle Hill community leader.

13. Justicia Ambiental, Desigualdad y Pobreza en Puerto Rico. Informe Multisectorial sobre las violaciones de derechos económicos, sociales, y medioambientales tras el paso de los huracanes Irma y María (2017) https://noticiasmicrojuris.files.wordpress.com/2018/05/final-informe-cidh-audiencia-pr-dic-2017.pdf.

14. Centro de Periodismo Investigativo, *Los muertos de María* (serie investigativa), September 2018, http://periodismoinvestigativo.com/2018/09/

los-muertos-de-maria/; Centro de Periodismo Investigativo, *Vidas de damnificados por María aún penden de un hilo*, September, 2018, http://periodismoinvestigativo.com/2018/06/vidas-de-damnificados-por-Maria -aun-penden-de-un-hilo/ and Centro de Periodismo Investigativo, *Investigación CPI+CNN: Puerto Rico tuvo un brote de leptospirosis tras el huracán María, pero el gobierno no lo dice*, July, 2018, http://periodismoinvestigativo .com/2018/07/puerto-rico-tuvo-un-brote-de-leptospirosis-tras-el-huracan -maria-pero-el-gobierno-no-lo-dice/.

15. ICADH, *Justicia Ambiental: Desigualdad y Pobreza en Puerto Rico,* 2017, https://noticiasmicrojuris.files.wordpress.com/2018/05/final-informe-ci-dh-audiencia-pr-dic-2017.pdf.

16. C. F. Weems, S. E. Watts, M. A. Marsee, L. K. Taylor, N. M. Costa, M. F. Cannon, and A. Pina, "The Psychological Impact of Hurricane Katrina: Contextual Differences in Psychological Symptoms, Social Support and Discrimination," *Behavior Research and Therapy* 45, no. 10 (October 2007), 2295–306.

17. Some residents fill the river channel with garbage and debris to construct their houses. In the past, a marshal was appointed to deter this practice. Some residents commented that to date, no security measures have been taken to finally stop this practice.

18. Jennifer Hinojosa and Edwin Meléndez, "Puerto Rican Exodus: One Year Since Hurricane María," Center for Puerto Rican Studies, September 2018, https://centropr.hunter.cuny.edu/research/data-center/research-briefs/puer-to-rican-exodus-one-year-hurricane-maria.

19. Ana Fernández, "Hacia los Estudios Transdisciplinarios de la Subjetividad: Reformulaciones académico-políticas de la diferencia." Revista Investigaciones en Psicología 1, nos. 0329–5893 (2011): 61–82.

20. Funded by Fundación Acceso a la Justicia.

AUTHENTICATING LOSS AND CONTESTING RECOVERY

FEMA and the Politics of Colonial Disaster Management

Sarah Molinari

The Federal Emergency Management Agency (FEMA) has consumed headlines in Puerto Rico since Hurricane María made landfall. From the prestorm understaffing and lack of Spanish-speaking workers to the undelivered pallets of water and big contracts for inexperienced bidders, news of the US disaster agency's problems grabbed local and international media attention. But the FEMA central operations hub had an air of efficiency and organization. The GRF Media building in Guaynabo that houses FEMA's central offices bustled like a newsroom on the day I visited to conduct an interview—rows of open desks, phones ringing, message boards, maps, and hundreds of staffers. Behind the appearance of smooth operations, however, were layers of failure and incomplete recovery. According to FEMA data,[1] the agency approved 464,821 applications for the Individual and Household Assistance Program, which covers repair costs to primary residence personal property and household structural damage in designated "essential living areas."[2] But this number fails to account for those who were approved but underfunded and for the more than three hundred thousand applicants who were denied. While many filed appeals, FEMA denied or did not respond to 79 percent of the appeal claims.[3] I argue that institutional failure gave way to contested recovery frameworks that challenge the assumptions of colonial disaster management and its disciplining tools.

UNEQUAL DISASTER AND "OTRAS MARÍAS" IN LAS CAROLINAS, CAGUAS

In the semirural neighborhood of Las Carolinas, Caguas, where I am conducting ethnographic research and collaborating with the neighborhood Centro de Apoyo Mutuo (CAM; Mutual Support Center), layered abandonment over time and demographic specificities exacerbated the impact of Hurricane María. sixty to seventy percent of the community's twenty-five hundred residents are over fifty-five years old, and one in three households lives below the poverty line.[4] According to the Residents' Association, Las Carolinas has lost 20 percent of its population since 2013, leaving behind elderly residents with tenuous family support. In the words of a CAM leader reflecting on the storm's one-year anniversary, "Before living through María, we experienced *otras Marías*. Because for the people with less resources, we're the ones who are always at the bottom."

Families left Las Carolinas in large numbers, especially in 2017 when the Puerto Rico Department of Education shut down the María Montañez Gómez elementary school as part of that year's wave of 167 school closures. Parents, residents, and teachers had struggled since 2012 to keep Las Carolinas's only elementary school open through protests, standardized test boycotts, strikes, and countless meetings with education and municipal officials. But the Department of Education pointed to below-average scores on standardized tests and declining enrollment, finally shutting the school down in spring 2017. Recalling the five-year struggle, one mother told me:

> My dad, aunts and uncles, my kids, and I all studied here. I remember the last sixth-grade graduation here was when my youngest son graduated. . . . I remember we held a strike to protest the school closing, and a friend of mine asked me, "What are you doing here?" And I said, "Why are you asking me this?" "Because

your son already graduated," she said. And I said, "What does that have to do with anything? My son graduated, but he came from here, and I still have his cousins here, and I'm here for them, and for your daughter too, who's still here.

After María, Las Carolinas residents went without water service for three months, without electricity for seven months, and without storm municipal debris pickup for eighty days. During the post-María wait for debris collection, some residents made a monthly ritual out of acknowledging the debris's "birthday" and decorating the piles lining the streets with Christmas ornaments. Trucks from the privately contracted company EC Waste finally appeared in early December 2017 after the Residents' Association president, Miguel Rosario Lozada, wrote an op-ed in *El Nuevo Día* critiquing the mayor and what he called "the absent municipal state."[5] Among other things, the municipality kept delaying debris pickup, barely circulated information, and originally scheduled Las Carolinas *last* among all the municipality's neighborhoods for power restoration.[6]

Even after debris removal, residents organized cleaning brigades to do basic vegetation maintenance like tree trimming and mowing around the public streets. The hollowing out of these municipal services certainly contributed to the posthurricane blackout and infrastructure collapse that was part of a longer process of public divestment. For example, the municipality of Caguas cut back on maintenance with a 14 percent total budget cut from 2016 to 2018, exacerbated by a $350 million cut from to the central government's contributions to all municipalities in fiscal year 2017–18—budgets, of course, approved and imposed by the colonial Fiscal Control Board, locally referred to as la Junta.[7]

Beyond the uneven effects of austerity and María, disaster assistance and recovery measures themselves may exacerbate vulnerability and inequality over time.[8] These contradictions have also shaped alternative ways of imagining recovery through *apoyo mutuo* (mutual support) rather than dominant individualizing and market-driven frameworks meant to restore private property, wealth, and the existing social order.

"TOTAL LOSS" AND THE STRUGGLE FOR A DIGNIFIED HOME

Jennifer, a resident of Las Carolinas, still lives with a blue tarp covering her roof fifteen months after Hurricane María.[9] She lives with five family members in the basement of her house because the top floor is uninhabitable. She applied for FEMA's Individual and Household Assistance Program, and FEMA inspectors arrived five weeks after María to declare Jennifer's house a "total loss." Jennifer has lived in her home for forty-four years, but like many others in the community who live on *parcelas* that have been divided over time and passed on through inheritance, she does not have formal property title.[10] In fact, a confidential FEMA memo from 2011 lists Las Carolinas as one of 116 areas in Puerto Rico that FEMA identifies as a "squatter community."[11] The memo provides inspection guidelines and states that squatters, and presumably the entire communities listed, are *ineligible* for home repair assistance because "they are not the owner of the damaged dwelling." In November 2018 the FEMA Puerto Rico News Desk confirmed with me via email correspondence that the 2011 squatter communities list had *not* been updated, although it remains unclear whether this list was used during Hurricane María and how the memo might have shaped prejudiced assumptions about "squatters" and identified communities. Legal activist groups like Ayuda Legal Huracán María pressured FEMA to revise its policy on verifying and assisting individual owners ten months after María.[12] Under the revised policy, applicants could submit a "declarative statement" to "prove" ownership with their FEMA appeals if they did not have a formal title, but this was too little too late. FEMA failed to systematically circulate information about this new option to verify ownership, leaving many who would have qualified out of the loop.

Because of her tenuous ownership status in the eyes of the state, Jennifer applied for FEMA individual assistance supported by an affidavit from a pro bono lawyer stating that she was the rightful owner and occupant of the home and that she was securing her formal title. To Jennifer's surprise, Puerto Rico's Departamento de Vivienda (Housing Department) finally began processing her application for a formal title after María, even though she had been applying for decades. She expressed to me a sense of relief because she almost had her title, as if this document might reduce her vulnerability to future storms.

FEMA installed a blue tarp on Jennifer's home and authorized $11,000 to repair structural damage, including a new cement roof, doors, and windows. Jennifer has not, however, been able to find a contractor to do the job within this budget because she estimates that $11,000 will cover only materials, wood, and some of the labor.[13] She appealed the FEMA decision to try to get a larger amount but was denied and could not understand how "total loss" amounted to $11,000: "They came and took photos. *They saw*," she said.

To fill in the assistance gap, the Departamento de Vivienda recommended that Jennifer apply for Tu Hogar Renace (Your Home Is Revived), a FEMA-funded program locally administered through the Departmento de Vivienda that provides up to $20,000 for minor emergency repairs to make houses "safe and functional." Tu Hogar Renace is organized around seven construction conglomerates (five US companies, two Puerto Rican companies) that have divided up the archipelago into contractor zones. Note that the Texas-based SLS Company (designated for zone 2) was awarded a $145 million contract in 2018 for Texas border wall construction, highlighting the entanglements between the US disaster and security industries. These seven major companies then contract smaller construction companies in Puerto Rico to provide labor and materials, but the major profits are funneled to the top. My interviews with Tu Hogar Renace and FEMA inspectors revealed the exploitative nature of inspection work. Inspectors working for FEMA's subcontractors described working seven days a week "de sol a sol" (from sunup to sundown) and driving in difficult conditions in their personal vehicles, often without GPS or phone signal, to inspect homes both before and after disaster aid was distributed. Inspectors, many of them low-paid Puerto Rican workers with little experience or formal training in this field, were generally paid per inspection, incentivizing them to work like "machines," as one inspector described. They act as the "eyes of FEMA" and collect evidence on electronic tablets using formulas, written comments, measurements, inventories, and photos to authenticate the owner's account of loss and damage; they then send the report to FEMA for quality-control reviews for final approval. Even though final aid determinations are out of their hands, inspectors described strategies like taking extra photos or photographing structural damage at dramatic angles to help residents receive the most aid possible. One of the main tasks for workers doing Tu Hogar Renace final inspections was to ensure that the house was "habitable" and that all construction materials and appliances were properly installed and made in the United States—if something was identified as an internationally made product, it failed inspection and needed to be reinstalled.

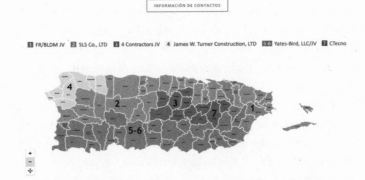

Tu Hogar Renace contractors. Screenshot of website, November 2018.

Tu Hogar Renace approved Jennifer for the maximum $20,000 for interior repairs like replacing the stove, sinks, refrigerator, cabinets, and beds and for installing electric generators.[14] But Jennifer kept delaying the work because it made no sense to install interior equipment without a proper roof. "I was waiting and waiting, time was passing, and the roof was becoming more damaged. And every time [Tu Hogar Renace] wanted to come and install the equipment, I had to stop them because without a roof everything inside the house would be damaged," she said. Tu Hogar Renace told Jennifer to call once she had the roof done. Then around September 2018, she called because she had gotten some temporary zinc panels on part of the roof, so she figured she could start installing the interior equipment little by little. But Tu Hogar Renace had closed her case without any notification. "Time was up," they told her, and the promised $20,000 vanished. Ironically, Tu Hogar Renace's webpage logo reads, "The first step to your recuperation" (https://tuhogarrenace.com/).

I asked Jennifer if she was going to appeal the Tu Hogar Renace decision, which she had a right to do, but she said no because "they don't answer the phones. There are always problems with the contact. And so here I am." After the slow violence of the bureaucratic marathon that seemed to work against her every step of the way, she felt resigned and overcome with a kind of disaster assistance fatigue. The slow violence of this wait for documents, inspections, follow-up calls, and contractors was not eventful or extraordinary but rather banal, everyday, and formative of Jennifer's experience of post-disaster trauma and the anxiety of living without a roof.[15] Jennifer's experience unsettles the normative temporality of disaster that moves linearly from response to recovery and rebuilding—without a proper roof, "recovery" seemed stuck in time for her.

DISCIPLINING DISASTER SUBJECTS

Jennifer's case highlights the disciplining role that property plays in dominant institutional disaster relief frameworks, or the tensions between FEMA's reliance on formal property title versus people's lived experience of diverse property relations in Puerto Rico. Discourses stigmatizing residents without formal property titles came up repeatedly in my interviews with officials from agencies like FEMA and COR3.[16] The project to "formalize property" is in fact central to official visions of "resiliency," mitigation, and disaster preparedness in part because it ensures the smooth operation of the homeowners' insurance market *and* future property speculation. The July 2018 draft central government Disaster Recovery Plan proposes $800 million to promote property title registration and establish penalties for those who do not register, but with little detail about how this will be executed. Furthermore, Puerto Rico's Departamento de Vivienda's amended Action Plan for federal CDBG-DR funds includes a proposal for a $400 million "Title Clearance Program" to formalize title for properties not located in floodplains or landslide risk areas.[17]

In late May 2018, I visited the Caguas Disaster Recuperation Center to inquire about municipality-specific FEMA data because I could not find this information online. There were only about four residents in a large gymnasium making their rounds through the information tables. In fact, there were more armed guards in the gymnasium than Caguas residents, although it was unclear what the security threat was. The director told me he could not give out any information, but he gave me a contact at the FEMA Media Relations Department. From there, I communicated via email and was vetted for over two months by both Media Relations and the News Desk to schedule an interview with a high-ranking FEMA Coordination Officer in August 2018. This process of accessing information and navigating contacts and offices was, of course, enabled by my own positionality as a white, English-dominant visiting researcher.

I was given a fifteen-minute time slot with this official and three of his staff members, who sat in the room keeping time and audio-recording the interview, which I was also audio-recording. I asked about FEMA's perspective on the widespread trouble people were having with their applications and appeals. Referring to FEMA's five- to ten-year vision, the official said:

> We're trying to change the culture of waiting for the last minute.... The people in Puerto Rico need to start thinking, "I need to take care of myself. I have to be responsible for my own future. I have to take action. I can't wait for stuff to be given to me." And I'm telling you because I'm Puerto Rican—this is cultural. They're waiting for stuff to happen. We need to change that to be proactive for our own future. If you're going to fix your house, don't do it halfway. Use the best materials.

I thought of Jennifer, who had been "taking action" all along, and all those for whom "recovery" is an ambiguous promise. The FEMA 2018–22 Strategic Plan echoes this culturalist argument—promoting a neoliberal "culture of preparedness," as the Coordination Officer described. The plan makes no mention of climate change, and its strategic goals include encouraging local jurisdictions to invest in predisaster mitigation through new financial instruments called resilience bonds; it also calls for expanding homeowners' flood insurance, incentivizing so-called positive behavior change through training and exercises, and increasing personal financial preparedness, which is hardly an easy task for vulnerable, racialized populations most at risk in these disasters.[18] For institutional agents of relief and recovery, the ideal "disaster subject" could be thought of as a private property-owning individual who has, a formal property title, home insurance, flood insurance, savings in the bank, and a ten-day supply of food and water for each person in the household.[19]

 FEMA ✔
FEMA @fema

Get financially prepared for emergencies. Consider saving money in an emergency savings account and be sure to keep a small amount of cash at home in a safe place since ATM's and credit cards may not work during a disaster. #FinancialFuture2018

FEMA tweet on emergency financial preparedness, April 9, 2018.

Personal indebting also figures as a disciplining tool of disaster recovery. In August 2018 representatives from FEMA and the

National Flood Insurance Program hosted a workshop in Las Carolinas to promote low-interest loans from the federal Small Business
Administration; these loans do not have the same use restrictions as
FEMA assistance and can be used for individual households or small
businesses to fill in relief funding gaps. After a long PowerPoint presentation, the FEMA representative assured a group of mostly elderly
attendees that these loans are secure and nondiscriminatory, unlike
private bank loans: "When we think about loans, we think about
Banco Popular, but this is a loan to *help* after a disaster," the representative emphasized. This was met by one frustrated resident insisting
that the representative review other disaster relief options because
debt is a hard thing to sell among low-income, elderly people.

REIMAGINING RECOVERY THROUGH APOYO MUTUO

But there are local alternative projects to reimagine disaster recovery
collectively rather than through racist cultural deficiency frameworks
and modes of indebting. For example, a number of autonomously
organized Centros de Apoyo Mutuo arose in the aftermath of Hurricane María, including one in Las Carolinas. Here, Jennifer and
other woman residents occupy the closed María Montañez Gómez
elementary school, transforming this abandoned space into a one of
healing, care work, and learning that has shifted in response to the
community's needs from the emergency to longer-term recovery and
local transformation. Women at the CAM cook and home-deliver
about one hundred lunches three times a week along two routes in
the community to elderly and bedridden residents and their care-
givers, providing a consistent presence and form of accompaniment.
This amounts to about six to eight hours of unpaid labor each day, in
addition to the incalculable emotional labor and their lives as mothers, grandmothers, partners, and daughters. Besides the lunch deliv-
ery, the CAM's major projects include a weekly acupuncture clinic,
gardens, a thrift shop, and an activity room for the elderly, where they
host talks and workshops, play dominoes, and do art projects. CAM
leaders are partnering with the Residents' Association to develop a
community emergency plan, which will include converting part of

the school into an emergency shelter and administering a census to identify residents' specific circumstances, like whether they take medications, live alone, and have limited mobility.

Centro de Apoyo Mutuo, Las Carolinas. Photo by author.

Local projects of *autogestión* cannot replace holding state institutions like FEMA accountable and demanding access to the disaster assistance and public services that people have every right to. But the politics of Apoyo Mutuo in this particular case demonstrate another path forward—albeit one constrained and shaped by relations of power—that is framed around support and affective relationalities.[20] This "social ecology of support"[21] rejects individualizing frameworks that render people as consumers in a recovery market who are disciplined through relations of property and debt and thus may point to a more equitable path forward.

1. FEMA, "Puerto Rico Hurricane María (DR-4339)," https://www.fema.gov/disaster/4339.

2. "Essential living areas" are spaces including the kitchen, bedroom, and bathroom. This category excludes spaces like patios, second or third bathrooms, or bedrooms above the home's number of residents.

3. Nicole Acevado, "FEMA Has Either Denied or Not Approved Most Appeals for Housing Aid in Puerto Rico," *NBC News*, July 17, 2018, https://www.nbcnews.com/storyline/puerto-rico-crisis/fema-has-either-denied-or-not-approved-most-appeals-housing-n891716.

4. American Community Survey, Bureau of the US Census, "Selected Economic Characteristics: 2013–17 American Community Survey Five Year Estimates," 2017, https://factfinder.census.gov/faces/nav/jsf/pages/community_facts.xhtml.

5. Miguel Ángel Rosario Lozada, "Un gobierno municipal ausente," *El Nuevo Día*, November 7, 2017, https://www.elnuevodia.com/opinion/columnas/ungobiernomunicipalausente-columna-2372408/.

6. After pressure from the *El Nuevo Día* column and a community protest at the local power servicing plant, Las Carolinas was reprioritized by the municipality and moved up in the power restoration schedule. Its power was restored in March 2018.

7. Caguas Municipal Budget, Fiscal Year 2017–18. https://caguas.gov.pr/wp-content/uploads/2017/07/Presupuesto-Modelo-2017-2018.pdf

8. See Junia Howell and James R Elliott, "Damages Done: The Longitudinal Impacts of Natural Hazards on Wealth Inequality in the United States," *Social Problems*, August 14, 2018, https://doi.org/10.1093/socpro/spy016; Vincanne Adams, *Markets of Sorrow, Labors of Faith: New Orleans in the Wake of Katrina* (Durham, NC: Duke University Press, 2013.

9. All names, except identifiable public figures, are pseudonyms.

10. In interviews and conversations, residents describe three main property divisions in Las Carolinas: *parcelas viejas*, *el fanguito*, and *la urbanización*. Parcelas viejas are plots of land distributed around the mid-twentieth century to those who "established" the community. Some parcelas have been subdivided among families or passed on across generations, and many do not have titles. El fanguito is described as *terreno invadido*, or squatted land. Residents of el fanguito may have been ineligible for FEMA aid because they live in a flood zone and were asked to relocate some years ago. Those who refused to relocate now do not have rights to FEMA assistance. La urbanización popularly refers to the more economically privileged residents, who all have titles to their property.

11. Federal Emergency Management Agency, "Memo on Non-traditional Forms of Housing," 2011.

12. Ayuda Legal Huracán María (ALHM), *Manual de abogacía para desastres Puerto Rico*, San Juan: ALHM, 2018.

13. As of the writing of this essay, Jennifer has kept the $11,000 untouched in a cooperative bank account and has refused to use it for other financial needs

that have arisen. She remains determined to use the allocated money to fix her house, even though it probably will not be enough.

14. Jennifer did not previously have electric generators, and I've found in multiple interviews that Tu Hogar Renace installs or adds things like smoke detectors and electric generators onto their jobs that were not there before or were not requested by the homeowner. These, of course, raise the aid estimate and thus the profits to the contracted companies.

15. On "slow violence" and the politics of waiting, see Javier Auyero, *Patients of the State: The Politics of Waiting in Argentina* (Durham, NC: Duke University Press, 2012); Rob Nixon, *Slow Violence and the Environmentalism of the Poor* (Cambridge, MA: Harvard University Press, 2011).

16. COR3 is the Central Office of Recovery, Recuperation, and Resiliency, an agency under the Puerto Rico Public-Private Partnerships Authority (P3).

17. The Community Development Block Grant Disaster Recovery Program is administered by the federal Department of Housing and Urban Development (HUD) and provides recovery funding to declared US disaster zones. The estimated CDBG-DR funds designated for Puerto Rico is about $20 billion.

18. Federal Emergency Management Agency, Strategic Plan, 2018–22, 2018. https://www.fema.gov/media-library-data/1533052524696b5137201a4614a de5e0129ef01cbf661/strat_plan.pdf.

19. See, for example, how FEMA's Strategic Plan, 2018–22 uses language on "resiliency" and building a "culture of preparedness."

20. *Autogestión* has no direct translation to English. It can be loosely translated to autonomous or self-organizing practices, often without any institutional or direct state ties.

21. Eric Klinenberg, *Heat Wave: A Social Autopsy of Disaster in Chicago* (Chicago: University of Chicago Press, 2015).

THE ENERGY UPRISING

A Community-Driven Search for Sustainability and Sovereignty in Puerto Rico

Arturo Massol-Deyá

Before Hurricane María struck Puerto Rico on September 20, 2017, I walked the streets of the small mountainous municipality of Adjuntas, in silence. I whispered my farewells with the strange feeling that everything would be different after the storm. Indeed, María's devastation of both the natural landscape and the built infrastructure became apparent everywhere. Hundreds of houses, many of them without roofs, showed significant damage. Landslides and fallen trees made roads inaccessible, and power lines were down everywhere. Bare trees made the landscape look near total desolation. There was no electricity, the potable water service was intermittent, and the only supermarket in town was destroyed by what looked like an act of war. Adjuntas was cut off from the rest of Puerto Rico and from the planet, and there was no fuel available. Hundreds of "refugees" were stuck in cramped, makeshift public shelters that had few cots to sleep on and that lacked food, medicine, and basic services.

The hurricane struck Puerto Rico after more than a decade of economic depression and austerity measures. As a result, government preparedness was every day more limited while resources were directed to a favored few. For example, the government used emergency decrees

298

to fast-track projects without having to abide by laws or regulations. The government did this in 2010 with an emergency energy decree to impose a ninety-three-mile-long natural gas pipeline through water resources, forestlands, agricultural areas, wetlands, and densely populated areas on an island one hundred miles in diameter. The government promised to solve the country's energy problems with a tube and thus spent millions of dollars—aggravating the fiscal crisis of the Puerto Rican Electric Power Authority (PREPA) instead of addressing Puerto Rico's real energy crisis: that the energy system is based on exploiting nonrenewable resources.

Given the well-documented effects of greenhouse gas emissions on global warming and climate change, pursuing a better future must unequivocally include addressing our dependence on fossil fuels. A group of researchers from Florida State University and Princeton University estimated that for every one-degree Celsius rise in the surface temperature of the sea, a hurricane's winds can accelerate by eighteen miles per hour, enough to change its category.[1] In addition, the increase in heat means higher rates of evaporation, or more water in the atmosphere. The higher temperatures produced Hurricane María, which grew to the size of France, carried a massive amount of water in its system, and thus had great destructive power, creating catastrophic floods and landslides. After María passed through Puerto Rico, we suffered the combined consequences of all these factors, which included over one hundred thousand landslides, as documented by satellite images and the US Forest Service. One of these landslides occurred in the municipality of Utuado, where two sisters were buried alive in their home. Others occurred in areas of steep hills where the ill-planned construction of roads and buildings heightened the risk of landslides.

To address these challenges and face the unique hazards of the twenty-first century, we must better comprehend the ecological role of forests in society, address water security, protect the coasts from sea level rise and storm surges, rethink how we practice agriculture, and redesign our homes to become more self-sustaining.

The problem is that we recognize the warnings but fail to take proper action. "Green" discourse now permeates politicians' rhetoric while in practice they perpetuate the burning of coal, insist on building incinerators, and open new valves for natural gas. Meanwhile, they impede those who promote renewable energy and the conservation of forests, which produce clean air, buffer against floods, and help prevent droughts.

The warnings about hurricanes and droughts have been on the table for quite some time. In a world of increasingly extreme weather, it is fundamental to promote a future that rejects crude capitalism, corruption, and the clientelism that pervades our energy system.

PUERTO RICO'S POWER CRISIS

Adjuntas lost its electric service days before the storm when it still had not seen an inch of rain or gusts of wind. No one was surprised when the hurricane left PREPA unserviceable and exposed. The storm laid bare PREPA's antiquated energy model, based on power plants that burn oil, gas, or coal. The plants' by-products—including ash and carbon air emissions—threaten the health of our people and our overheated planet. To this we can add a huge public debt that represents one more aggression against our children, who will end up migrating or paying for the mediocrity of corrupt past decisions.

Simply restoring the same failing infrastructure is not progress. The country must go from mere aspiration to constructing a responsible and resilient energy model based on renewable-energy sources, such as water, wind, biomass, and, of course, solar power. Yet, with only a small window of time to advance alternative sources of energy that might forestall climate change, PREPA remains obsessed with previously discarded options, such as natural gas, that should be left in the past. No more than 3 percent of Puerto Rico's electricity is generated with renewable-energy sources, yet the government affirms that the system cannot increase this percentage because the energy system "was not designed for renewables" and because "the sun shines only during the day."

One thing remains clear: we must abandon fossil fuels and PREPA's obsolete energy model. PREPA maintains its astronomically high prices because it monopolizes the energy market and blocks efforts to pursue alternative energy sources.

PRIVATIZATION COMPROMISES PUERTO RICO'S ENERGY DEVELOPMENT

If there is something we know well by now in Puerto Rico, it is the integral importance of our energy system. The hard lesson was learned, unfortunately, because of the system's obsolescence. After Hurricane María, we counted the time without service in months, not days. The absence of electricity caused greater insecurity, unemployment, bankruptcies, water contamination, as treatment plants discharged raw sewage. All this was exacerbated by hospitals' inability to operate at full capacity. In the six months following the storm, 2,975 people died and many more fled to the United States.

With a population overwhelmed and suffering, many received the government's announcement to privatize the country's electric system as the logical step forward. The government claims that the power company's failures are a result of its being publicly owned and that it is thus the victim of partisanship. After María, however, Puerto Rico was failed by both public and privatized services (such as telecommunications). Yet the public energy sector is decried, while private telecommunication companies are given a pass. Undiscussed is that 30 percent of PREPA's energy generation was privatized in the mid-1990s. Privatization is not a new solution. The privatized component of PREPA has largely dictated the company's contracts and new projects, which tripled the public debt even as its ability to operate was decimated by the imposition of austerity measures.

The private model has its sights set on expanding PREPA's use of natural gas, coal, and incinerators. This vision of Puerto Rico's energy future, clearly expressed by the governing Fiscal Control Board, is the greatest challenge facing the country. Instead of maintaining our dependence on fossil fuels and spending over $2.8 million annually just for fuel, we must aim for energy self-sufficiency.

Our experience after Hurricane María should be sufficient to instigate a change in direction. Since the storm, PREPA has resorted to subcontracting recovery operations to private, mostly US-based companies. There is no longer a local-public potential to respond because it had been dismantled. Fiscal austerity measures had led to the elimination of maintenance brigades and response bodies, such as the Defensa Civil which used to be in each municipality ready to respond immediately after the passage of a hurricane or storm. Corrupt schemes were uncovered, such as the contracting of the infamous Whitefish company, yet after that, similar examples followed. Repair workers brought in from outside the island were up to ten times more expensive than local union labor, and after months of contracts, much of the country still had no power. It took almost a year for PREPA to announce that power had been fully restored to Puerto Rico. Even still, of course, they left many in the countryside in the dark, since it was unprofitable to provide them with electricity unless the consumers themselves paid for utility poles and assumed other costs.

Initially, the number of fallen utility poles was used to define the degree of damage. According to estimates, there were sixty-two thousand fallen poles (although this figure was later reduced to forty-eight thousand). Based on this and other observations, it was declared that 80 percent of our energy system had collapsed. Yet none of the power plants suffered significant direct damage from the hurricane, and in actuality over six hundred thousand utility poles remained unaffected. Surely that number of fallen poles is substantial, but it did not represent even 10 percent of the total. The damage to the electrical system has been overstated as a means to conceal the incompetent restoration process. Surprisingly, the privatized repair effort did not expeditiously fulfill its task.

IF THE VÍA VERDE GAS PIPELINE EXISTED TODAY

In 2010, Governor Luis Fortuño's administration declared a state of energy emergency. The solution, according to them, was to build a ninety-three-mile-long gas pipeline. The pipeline, deceptively called La Vía Verde (the Greenway), would go through the steepest slopes of La Cordillera Central, pass over (or below) 234 bodies of water, fragment farms and forests, and pass through residential courtyards and over wetlands to reach PREPA's generating units in Palo Seco and San Juan.

What would Hurricane María's impact have been if we had allowed the construction of Vía Verde? Would the electricity have been restored faster? Would thousands of deaths have been prevented by having a gas pipeline? Would the emissions and shrill noises of power plants have been avoided? Of course not. We can surmise, however, that the landslides would have been worse. The pipeline would have passed through exactly the mountainous areas where landslides dragged down homes, bridges, roads, and other infrastructure. Without trees and their roots, soils are more prone to landslides.

It was gross negligence to waste time and resources on the pipeline project, which went nowhere, while neglecting the energy system's critical needs. The government knew the pipeline was technically infeasible because Ecoeléctrica, the natural gas terminal in Peñuelas, lacked the capacity to process enough surplus fuel to supply the pipeline. Moreover, trying to sell us a pipeline as the solution to our energy crisis was an unjustified and corrupt act. Instead of maintaining and strengthening the transmission and distribution lines, the government dismembered PREPA with its austerity, waste, and bad decisions. As a result, PREPA went bankrupt.

Today's posthurricane marauders, politicians and gas and oil industry representatives, lobbying behind the people's backs and making decisions about the country's reconstruction, are the same ones who wanted to distribute pipeline contracts among their friends. In confronting them, we must maintain an energy agenda that can be summed up with the words "renewable resources." It's the only way to counter those whose farcical discourse calls a pipeline a greenway.

IMPOSING ENERGY SLAVERY

Representative Rob Bishop (R-Utah), chairman of the House Committee on Natural Resources, visited the island in May 2018 in the middle of the extended emergency. "I'd like to see more natural gas ports here," Bishop said, admitting that he discussed the issue in Washington, DC, with executives from private companies that he declined to name. His energy agenda for Puerto Rico excludes solar and wind power, condemns the Caribbean to dependence on fossil fuels, and uses Puerto Rico as both a consumer and a distribution hub for the region.

Deliberating and making decisions behind our backs, Washington politicians (influenced by powerful lobbyists such as former governor Fortuño) see a country to rebuild in a way that suits their interests. The US government wants to keep us subjected to a model of energy dependence. That is, "energy must be imported," Bishop said, and "natural gas is a brilliant way to do it." The new American fever is not Californian gold or the oil of the 1960s, but fracking, or hydrofracturing—one of this century's most destructive technologies, for extracting underground gas reserves. It is therefore wrong to think that the United States has lost geographic interest in our islands and would prefer that we become either a US state or an independent nation. The energy agenda and the privatization model clearly indicate US aims: to perpetuate Puerto Rico's colonial status quo.

Faced with this unfortunate and unjust colonial reality, we must organize locally to reject the imposition of US energy dependence and propose an agenda of energy self-sufficiency—one that emphasizes the clean and safe energy sources that we have in great abundance. Generating energy with the sun, water, wind, and biomass through micronetworks, hybrid systems, and other configurations at the point of consumption is a route to achieve welfare and progress for all and to start to decolonize Puerto Rico.

TO GET OUT OF THE DARKNESS, BUILD AN ALTERNATIVE ENERGY PRESENT

Surviving the enraged force of a modern hurricane, and then suffering the collective despair of seeing so many people die because of the US government's historic gross negligence—this would make anyone seriously question the immediate future.

But in Adjuntas, we showed that alongside the despair there exists another Puerto Rico, one where diverse, organized communities responded to the emergency with self-management and a collective sense of hope.

Casa Pueblo, a local community organization, served as an energy oasis for the community, having dedicated itself to operate with renewable energy since 1999. Casa Pueblo's newly modernized system of solar energy resisted the hurricane's attack, allowing the community to undertake humanitarian aid tasks in the village, rural beltways, and other municipalities. As one of the first mitigation tasks, the #iLuminarPRconSOL initiative sought to mobilize the Puerto Rican diaspora and friends of our self-management project to overcome the collapse of our electrical grid. As a result, thousands of solar lamps illuminated family homes throughout large areas.

Given the collapse of Puerto Rico's communication systems, it was crucial in the first days of the crisis to connect hundreds of *adjunteños* with their families through satellite telephony, community radio, and our social networks. Meanwhile, multiple community chainsaws were passed from hand to hand to help clear roads and residences. Volunteers distributed coolers, water filters, clean water, food, and hygiene products. We also collected tarps from the diaspora, which were brought directly to Puerto Rico by volunteers when mail services were blocked. We distributed them in the mountains weeks before FEMA even arrived.

We decided to broaden our response with the goal of changing the energy landscape through energy self-sufficiency projects that would address the right to energy, communications, health, nutrition, entertainment, education, and economic stimulus.

Since February 2018, for example, the Pérez Barber Shop in Adjuntas has been operating with a solar energy system that Casa Pueblo installed as part of its economic development initiatives. Don Wilfredo, the barber, is the first person I have known whose eyes sparkle when he is asked about his electricity bill. Instead of the $75 to $80 that he used to pay monthly, his new cost for solar power is about $5.80.

One day, he told me, "I didn't realize there was a blackout. The customers are the ones who tell me when there's no light in town." He said this with a combination of pride, happiness, and empathy for those who remain in need. "There are many without power in the countryside," he said.

The barbershop joins a portfolio of Casa Pueblo's energy projects at the point of consumption that work directly with people and businesses. In just a few months after María, we had equipped ten homes in the rural community of El Hoyo with backup systems for medical equipment such as dialysis and respiratory machines. We addressed food safety with over fifty solar refrigerators installed in residences in all neighborhoods throughout the municipality, and five grocery stores now operate with solar energy. The groceries, which represent the first line of food defense for the community, are strategically located in the neighborhoods of Guilarte, La Olimpia, Vegas Abajo, Tanamá, and Garzas. In terms of education and entertainment, a solar lounge was established in the Bosque Escuela, and the first solar cinema in Puerto Rico was completed in April 2018. We also established the first fully solar radio transmitter and provided solar power to two hardware stores, sixty homes (which neighbors call the *cucubanos*, or fireflies), houses with peritoneal dialysis needs, a young worker's first *lechonera* (restaurant that serves roasted pork), and even a pizzeria, El Campo es Leña.

The hardware stores are expected to pass on their energy bill savings onto the community by offering reduced-priced, energy-efficient products for their customers. Supermarkets are asked to control their prices, not to close their operations, to offer better-quality food, and to serve as energy oases in the same way that Casa Pueblo has done for our people. Those who received solar refrigerators from Casa Pueblo promised to help their neighbors, and the barber promised to lower his rates. This is a model of change that leverages new resources to promote collective social responsibility.

Meanwhile, we have been working with professors and students from the University of Michigan on innovation to diversify energy sources and develop hybrid systems for an energy microgrid. The plan? To integrate agricultural biomass with vegetative residues to generate hydrogen at night and combine it with a daytime photovoltaic system.

IN SEARCH OF ENERGY INDEPENDENCE FOR PUERTO RICO

At a time when the central government lacks a clear vision for the future or it is being driven by external agendas, Casa Pueblo and many other groups and communities are advocating solar-power projects throughout the island. Where consensus seems to emerge is on maximizing local energy resources and on replacing an outdated model in which, shamefully, only 0.41 percent of the energy generated is solar. As we move forward, we are advocating a short-term goal of "50con-SOL" by 2027. That is, we want 50 percent of the country's energy generation for residential consumption to move to photovoltaic power by the tenth anniversary of Hurricane María. This would be the first phase in establishing a 100 percent renewable-energy system. In advocating this goal, we are also promoting an economic model that creates savings and benefits for local communities while creating jobs across the island. Furthermore, these initiatives will build resilience while promoting economic revitalization by empowering marginalized communities in Puerto Rico through local control of resources and creating a path to become a clean energy hub in the Caribbean. Finally, energy self-sufficiency could be our first step toward decolonization.

1. James B. Elsner, Sarah E. Strazzo, Thomas H. Jagger, Timothy Larow, and Ming Zhao, "Sensitivity of Limiting Hurricane Intensity to SST in the Atlantic from Observations and GCMs," *Journal of Climate* 26 (August 15, 2013): 5949–5957, https://doi.org/10.1175/JCLI-D-12-00433.1.

COMMUNITY KITCHENS

An Emerging Movement?

Giovanni Roberto

I work at Puerto Rico's *comedores sociales* (community kitchens), a social project where I cook, help with organizational tasks, and promote political activism. The name comedores sociales comes from the donation-based food stands that we started in 2013 at the University of Puerto Rico, Río Piedras campus, and other locations throughout Puerto Rico. The group that started the comedores is socialist, so we critique the capitalist system and try to create transformative experiences for social change, which include personal experiences for everyone involved in the project.

In my case, food and hunger have always been an issue. For many years my family had to resort to food stamps, which, as we know, help but are not enough. To have some cash income we sold pigeon peas, *ajíes*, pumpkins, mangoes, and anything that grew in the nearby mountainside. My mom would make *pasteles* and other things. My dad worked sporadically until his back was permanently injured and his employment options were even more limited. In our home, unemployment and self-employment were the norm.

My political awakening occurred with the historic events of the late 1990s, including the Huelga del Pueblo (People's Strike) of 1998, the release of Puerto Rican political prisoners in 1999, and the intense struggle against the US Navy's presence on the island of Vieques.[1] These were *historic, collective* events I experienced in my *individual context* of being from Puerto Rico's *montaña* region, of being hungry, and of living in poverty.

This is why I was forever convinced that we must construct different social, political, and economic systems, ones that would replace capitalism and all its practices of exclusion.[2] In Puerto Rico global capitalism manifests itself through the direct political control that the United States exercises over us in Congress. The capitalism that oppresses us here is a colonial system, and this is why our struggle for a better life is always *independentista* (proindependence) and anticolonial.

This kind of domination often leads to ideological musings like "What would we be without the United States?," "We are better off than any other Latin American republic," and "If we become independent, will we die of hunger?"

A PERSONAL JOURNEY

At the end of 2012, I was depressed and disoriented. My mood was perturbed, and I needed to reinvent myself. Quitting my teaching job at an alternative school was simple in comparison to deciding what to do with my life.

I eventually found the psychiatrist Efren Ramírez, who takes a holistic approach to mental illness. Even though he can prescribe pharmaceutical medications, he bases his practice on nutrition, including the consumption of natural lithium as a supplement to the process of personal development. Ramírez defended the idea that life can be lived through "mutual help and self-effort" in the search for meaning, which is possible with organization, nutrition, and community service, among other things.

His method made sense to me, and I immediately started to improve. I started making different decisions, and they had a ripple effect. At the sessions with Ramírez, I would share only the things I was doing to feel better, not my problems or things that made me sad. It was a way to focus on the search for solutions and to implement alternatives to meet life's challenges.

My mom, Guillermina Caez, who is also a founder of comedores sociales and one of its longtime cooks, was diagnosed with cancer in 2012. Her treatment was luckily simple and short because it was detected early. Her illness and recuperation influenced her diet and mine. Along with Ramírez's healing program, her illness made me rethink the role of food in human physical and social development.

The first food tables I set up, in 2013, were meant to accompany the political work I was doing with the International Socialist Organization (ISO);[3] I established the comedor from the perspective of organizing, which was what most interested me as an activist. To my surprise, in less than a year the food tables came to *substitute for the political organizing*, confronting us with another dimension of political work that we hadn't thought about before. It also presented me with a dimension of my purpose in life, one that I hadn't integrated into my political practice: the struggle against hunger and scarcity.

POLITICAL AND COLLECTIVE SOLIDARITY

The first tables sold a meal for four dollars or whatever people could manage. We also accepted donations and help. Because of the life I have lived, I dislike how money limits people. When distributing food, I didn't want money to be a limitation. Since the work I did at the tables was done with my mom's help from the beginning, receiving help with the tables was very important. If someone donated some rice or a can of beans, it was greatly appreciated, and we gave them a plate of food. Little by little, through our interactions with people, we developed our "system of three contributions": in exchange for meals, people can give the comedor money, a donation in kind, or volunteer labor. The starting point was always an antisystemic political perspective.

Tensions quickly ran high, and the university made it clear just how radical our actions were. The authorities of UPR Cayey tried to stop us, at the request of the manager of Fazaa Food Service, the private administrators of the university cafeteria, various times without success. Thus, the state and a private enterprise joined forces to oppose a food stand that gave away at most twenty meals a week to students who could barely afford a monetary donation. In time, I was issued a fine, but our accountant was able to get it removed. On another occasion the campus security tried, unsuccessfully, to force us to stop giving out the meals.

At the UPR Río Piedras campus, campus security guards once blocked the food tables with their bodies. Now it was the state practicing a *peaceful protest* against a civilian student initiative. The security guards, most of whom were obviously embarrassed, stood there to stop the distribution of the solidarity meals. After a while the rector at that time ordered the guards to leave. On another occasion the dean of the School of Business Administration cut off access to the water supply. And every now and then a new, cocky security guard would block the entrance for the comedores' car to deliver the food that was to be served that day. But unlike at UPR Cayey, the administration at UPR Río Piedras has always been open to regulating our project.

We at the ISO undertook the comedores sociales as part of our political work while also establishing the Centro para el Desarrollo Politico, Educativo y Cultural (Center for Political, Educational, and Cultural Development),[4] a nonprofit that we created in 2012 to promote experiences of change. In time, there grew among us a division between those who were undertaking the comedor and the ISO's student members who were oriented toward protest and movements.

Every attack from the administrators or the UPR strengthened our protest with food tables and gave us political experience. This is how we took on the name comedores sociales, and we explained the project's objectives in writing to the UPR authorities: we fed solidarity as we built a nonprofit organization dedicated to constructing long-term experiences of change for people. When we were questioned about our "business," we responded that it was in sync with the model of nonprofits: we accepted donations, materials, and volunteer labor. In any case, if the person could not provide any of the three types of donation, they ate anyway, making our project more one of free distribution. Yet we organize around solidarity, not charity. This is why we established flexible donation methods, including receiving a meal without a donation. This allows us to present a relationship distinct from that of the traditional customer-seller. We are in effect meal providers but with the goal of consolidating social ties and building community. The funds we receive pay rent and other operational expenses, including for personnel who work on the project.

GROWTH IN THE CRISIS

The past years have been particularly difficult for political movements that look to improve people's living conditions—difficult for the proindependence, socialist, and labor movements. Framed by the global crisis and debt, the upper class is waging a head-on war with the working class. The principal effects are obvious: the destruction of public service, the increase and consolidation of unemployment, and a high levels of migration. No one seems to be able to escape these effects.

Sadly, our comedores sociales are serving a lot of people because hunger grows in parallel to the crisis. A person who came to eat told me that he eats lunch at the comedores, goes to a Christian soup kitchen in the late afternoon for dinner, and asks his neighbor for food every now and then to survive the week. Someone else told me about the university students, "People think they are going hungry very often. You go to their fridge in the dorms, and there is absolutely nothing there." This is more common than one would think.

Comedores sociales are expanding around the island. There are now comedores at UPR's campuses in Ponce, Arecibo, and Humacao, as well as at the Universidad de Sagrado Corazón. In December 2015, we held the first island-wide meeting of the comedores sociales. In 2017, just when UPR Mayagüez was ready to start theirs, Hurricane María struck.

MUTUAL AID AFTER HURRICANE MARÍA

Comedores sociales had always been a precarious initiative, so when María hit Puerto Rico, all we had was food. What luck! We had collected provisions at the beginning of September to "have a strong start to the semester," and these provisions became the initial food for the *comedor social comunitario*, as part of the Centro de Apoyo Mutuo (CAM), which we founded together with the collective Urbe a Pie[5] and an army of volunteers, to address food scarcity in the aftermath of the hurricane. We used the "three donations" model to invite people to "not only receive a meal but to construct something long term." Since we knew that after a natural and social disaster, "disaster capitalism" would follow, and with it the attempts of the system to reestablish itself, we wanted our initiative to have as little governmental intervention as possible.

In this way, the CAM counted on the support of the very same people in the food lines and our fellow Puerto Ricans living abroad who were quick to send nonperishable food, medical supplies, electric generators, and solar light bulbs, among other things.

Somehow the hurricanes made the government's negligence clear—the well-being of the general population wasn't the government's priority, but the system of riches and wealth surely was. Even with the ports loaded with gasoline and merchandise, the government limited its distribution. During the first weeks, hundreds of truckers offered to work for free with no success. Since money was the motivation, not quickly reestablishing national electric power, the government waited and handed out shady contracts costing millions of dollars to the Whitefish company. Our government hid the number of deaths as a consequence of Hurricanes Irma and María to impress the US government while they were leading a hidden campaign in favor of annexing Puerto Rico to the US. While people were dying for lack of medical attention, our politicians were scrambling for loans, contracts, and political positions.

During the days right after María hit, and in the weeks and months to follow, it was the dozens of brigades, the comedores sociales, the independent medical efforts, the artistic presentations, and help from people in our country and from abroad that lifted up Puerto Rico.

The idea of *centro de apoyo mutuo* was embraced by different sectors in Puerto Rico that established similar centers in various towns. From Las Marias and Lares to Humacao and Vieques, different initiatives called themselves *centros de apoyo mutuo* to insist on a kind of dependency and assistance that is promoted by the state and its agencies. The call to construct something else keeps opening the way to more grassroots work with people.

This is how the comedores sociales found themselves involved in initiating other social community kitchens, like the one in Yabucoa with the support of the Centro de Desarrollo Político, Educativo y Cultural (CDPEC) and the Centro de Apoyo de Las Carolinas, a community in Caguas that has provided meals and other services to its residents since 2017. To strengthen relationships among each of the centros de apoyo and other similar initiatives, we created a network called the Red de Apoyo Mutuo (Mutual Support Network), which continues to do important grassroots work.

AN EMERGING POLITICAL MOVEMENT

Our comedores sociales model sustainable mutual aid in the area of feeding people, and the model is growing to become applicable for work in many other areas. A diversity of centers oriented toward satisfying social needs have been started in Puerto Rico over the last years, even before Hurricane María.

There is the Centro de Estudio Transdisciplinario para la Agroecología (Transdisciplinary Study Center for Agroecology) in Lares and the theatrical initiative of a home-based workshop Colectivo Columpio in the town of Camuy. There is also the Consejo Integral Comunitario de la Barriada Morales, which empowers and assists the community residents.

Ecological and agricultural projects are very prominent within an ecosystem of concrete alternatives. The return to agriculture and organic farming is the rising social tendency most important to consider for a political change in Puerto Rico over the next decade.

More than a movement, the comedores sociales are emerging in Puerto Rico as a political movement orbiting around satisfying basic needs. These initiatives can multiply because they respond to concrete necessities and are very easily reproduced. These initiatives are sustained by their own resources, private donations, or literally the passing of a hat. Instead of depending on the state, the federal government, or corporations, these initiatives "depend" on their own people. The direct connection to the *people at the base* is the calling card of these social and political projects directed at daily life. The crisis has proved to be an opportunity.

Sooner or later the crisis would have generated a collective response much like ours in the area of people's need for meals, as it would also generate other responses we don't know of. The crisis will make the initiatives and actions that respond to the needs of the people successful.

For the vigorous development of antisystemic politics in Puerto Rico, we need to be anchored in social projects just as the Black Panthers[6] and the Young Lords were in the United States. These organizations protested the government for its systematic abuse, but they also organized the resistance to capitalism every day.

These projects increase collective confidence and effectively demonstrate that we can do things better at the base. The work done to increase self-esteem, transmit confidence, and project hope can't be underestimated in Puerto Rico, because we are a country whose collective psyche has been pounded on by colonialism.

Our social projects have the virtue of creating new experiences of exchange and of human relations. They can help us overcome the fear about our survival and our political future sown into us generation after generation. "With independence we will die of hunger" is what people think and say, but with colonial capitalism we are dying now. An antiestablishment political movement has to be developed—one whose objective is to organize with the people at the base to concretize alternatives to the crisis. We need to fight the old world and construct a new one.

It works. One day at the comedor social, a guest asked how much the meal cost, and we responded, "The suggested donation is five dollars, but you can give what you can." She laughed at first, and then we noticed that it was from the surprise. She insistently asked, "But how much do I have to give?" And we said five dollars if you have it, to which she responded, "What do you mean 'if I have it'?" If I don't have it, then I don't have to pay?" We made it clear that if she ever didn't have money and needed to eat, she could come and help in another way, such as by giving a hand or doing some dishes. She gave us the five dollars and walked away. She kept coming the whole semester.

We don't have to ask about someone's politics because the comedor itself is a filter. If the person isn't solidary and does not want the best for others, we probably don't know each other. If they don't believe that people organizing can make changes, they probably won't trust our new projects because these projects do not operate under the existing authorities, and so they don't consider us legitimate. Profound change takes time.

This is why I have faith in future struggles. Generating change means generating a movement for modifying the paradigms of political functions. This is how our projects are simultaneously nourished by the crisis and are radicalized as a result of it.

Even though young people and women stand out, these projects of daily life integrate all kinds of people. Above all, the participation of the elderly is vital and indispensable in agroecological projects. All share the strong connection between the issues of health, culture, and general well-being.

The movement that we need to put into action, a project of national liberation, will be enriched by a combination of political struggle against the state and the short-term construction of concrete alternatives. We are doing it. We have to keep sowing.

1. The Huelga del Pueblo of 1998 was a series of protests, walkouts, and strikes against the privatization of the national phone service. It included a two-day general strike in all of Puerto Rico during the summer. Twelve political prisoners were released after nineteen years of imprisonment after an intense campaign and popular support. In 2010 Carlos Alberto Torres was released, and Oscar López Rivera in 2017. Vieques is an island municipality of Puerto Rico, invaded and occupied by the US Navy since 1945. After the death of a civilian, David Sanes, during military practices in 1999, the protest against the navy demanded an end to the bombing practices and the closing of the naval base. The struggle expanded until the navy base finally closed in 2003.

2. I was a militant member of the Organización Socialista Internacional (OSI) from 2000 until its end in 2014. This experience gave way to the creation of Centro para el Desarrollo Político, Educativo y Cultura (CDPEC), the Comedores Sociales de Puerto Rico, and other projects.

3. The ISO was a Marxist organization and defender of "socialism from below," the idea that an organized working class can and should run society. In 2013 there were two main branches, in Cayey and Río Piedras.

4. Comedores Sociales de Puerto Rico was initially a Project of CDPEC. For more information, visit www.cdpecpr.org.

5. Urbe a Pie is a collective that works for the social and economic development of the traditional town area of Caguas. They have a variety projects, such as Huerto Feliz, Boutique Comunitaria, Galería de Arte Comunitario, and Café Teatro El Reflejo.

6. The Black Panther Party was a Black nationalist, socialist, and revolutionary organization in the United States from 1966 to 1982. Its militants were involved in "survival programs" that served breakfast to children, gave clothes in the winter, and kept watch in the barrios for white police.

BUILDING ACCOUNTABILITY AND SECURE FUTURES

An Interview with
Mari Mari Narváez

Marisol LeBrón

Mari Mari Narváez is the founder and executive director of Kilómetro 0 (Kilometer Zero), an organization that fights state repression and violence against citizens. Growing out of the work of the citizen accountability group Espacios Abiertos (Open Spaces), which was established in 2014, Kilómetro 0 has been at the forefront of the struggle to make the state, especially the Puerto Rico Police Department, more transparent and accountable to the people of Puerto Rico. In particular, Kilómetro 0 has been dedicated to reducing instances of police harassment, violence, and lethality, while working with the public to gather stories of police misconduct and increase awareness about people's rights when confronted by the police.

Mari Narváez has been a fixture of progressive politics in Puerto Rico for more than two decades as a journalist and social commentator for a range of publications, including *Claridad*, *El Nuevo Día*, and *80 Grados*. Along with Sofía Irene Cardona, Ana Teresa Pérez Leroux, and Vanessa Vilches Norat, she published the books *Del desorden habitual de las cosas* (Capicúa, 2015) and *Fuera del quicio* (Santillana, 2008). She is also the coauthor of *Palabras en libertad: Entrevistas a los ex-prisioneros políticos puertorriqueños* (Editorial Claridad, 2000).

In this interview, she discusses how she came to political consciousness and began organizing around human rights in Puerto Rico. She discusses the political landscape before Hurricane María and how protests against the imposition of PROMESA and the Fiscal Control Board were met with intense state surveillance and police repression. For Mari Mari Narváez, Hurricane María and its aftershocks present new challenges and opportunities for social justice advocates and organizers in Puerto Rico. While Hurricane María and the storm of willful neglect and austerity that followed created new urgencies in the lives of many Puerto Ricans, which often makes taking to the streets to protest difficult, she points out that the storm has also ripped away the facade of democratic governance and revealed the realities of colonialism in Puerto Rico that can spur renewed political consciousness and organizing. Ultimately, Narváez suggests that a true recovery for Puerto Ricans living with the aftershocks of Hurricane María means building community capacities and working toward community control of the institutions that govern our lives.

MARISOL LEBRÓN. As a journalist and an activist, you've long been concerned with social justice issues and the struggle for human rights in Puerto Rico. Can you tell us a little bit about your political journey and how you became involved with efforts to create greater police accountability?

MARI MARI NARVÁEZ. Although I was raised in a family of activists, I never really meant to be one. I had wanted to be a journalist and a writer since I was an adolescent. Writing was and still is my greatest love. But when you think of about it, what are journalism and writing if not forms of activism, right? I began my journalistic career in *Claridad*, when I was twenty years old. A few years later, I was writing opinion columns in both *Claridad* and *El Nuevo Día* as well as working for different nonprofit organizations in Puerto Rico. So, writing about so many urgent social and cultural issues, I guess I slowly and unwillingly became an activist. I still struggle with it because I envisioned myself writing fiction stories and living a more peaceful life. Of course, that's just not going to happen ever, apparently. The times we are living are far more complicated than I ever expected.

My most life-changing political awakening was relatively late in life, in my late twenties. It happened in 2005, when the FBI arrived in Puerto Rico, formed a perimeter around Filiberto Ojeda Ríos's house, and killed him. Of course, I was already politically conscious and had long worked as a journalist. I had grown up between protests in the proindependence movement. But something happened that day, and it had to do with a vital and productive outrage. The human rights struggle got very personal for me. Filiberto's assassination changed me radically. Whatever I didn't fully "get" before about our political and colonial condition, I understood from that day on. And I knew I had to be more politically involved. Writing was powerful and efficient, but it was not enough. It was since that day that I felt the fire burning, and I still do. With time, things in Puerto Rico have only gotten worse from a colonial and socioeconomic stance, so my rage since September 23, 2005, has proved fruitful. I also had the privilege of working with great nonprofit organizations from which I learned and gathered so much. One thing led to another.

MARISOL LEBRÓN. There is a long history of the police working as a tool of political repression in Puerto Rico, a history with which you are intimately familiar. How do you understand the role of the police in Puerto Rico?

MARI MARI NARVÁEZ. My family has been ideologically persecuted by the police most of our lives. Before I was born, in 1976, my brother Santiago Mari Pesquera was killed in a political assassination in San Juan. A group of extremist Cuban exiles was involved in the crime's planning and execution, and the Puerto Rico Police participated, at the very least, in the cover-up. More than thirty years later, declassified documents from the FBI revealed that Cuban exiles had a plan to kill my father and the FBI knew it. Back then, my father, Juan Mari Brás, was a prominent political leader in Puerto

Rico running for governor as the head of Partido Socialista Puertorriqueño, an anticolonial, socialist, and secessionist movement. Five months later, my brother was killed.

Of course, my family is not the only one whose members were killed, persecuted, incarcerated, and blacklisted. In the 1980s civil rights litigation revealed that the police had illegally created political files to surveil dissidents, affecting the lives, jobs, and relations of anyone who protested or was even minimally involved in any progressive movement in Puerto Rico. Hundreds of people have been killed and thousands have been incarcerated or persecuted because of political repression and colonialism in Puerto Rico.

When I was twenty years old and became a journalist, I again encountered the police, this time in the social protests I was covering, where they would abuse protesters and journalists alike with excessive force, by intimidating people and repeatedly using tear gas against us.

I have no doubt that self-determination, as well as political and social progress in Puerto Rico, has been stalled, mainly, by the systematic use of repression. Specifically, police repression. But this not only applies to activists. It applies to underserved communities, which sometimes in a lifetime might not receive decent basic services from the state, but do receive the *mano dura* (iron fist), the discriminatory vigilance, the targeted abuse, and human rights violations from the police. This also pushes back social progress. These are some of the reasons why, since 2014, I have been dedicated to learning about, investigating, pursuing, and advocating for police accountability in Puerto Rico, first from Espacios Abiertos, a nonprofit in Puerto Rico, and now from Kilómetro 0, an organization I recently founded with their support.

MARISOL LEBRÓN. Two of the organizations you've worked with in a leadership capacity, Espacios Abiertos and Kilómetro 0, have had transparency and access to information at the core of their mission. Why is this so central to the work of transforming the relationship between the police and the people of Puerto Rico?

MARI MARI NARVÁEZ. The Puerto Rican government is and has

always been extremely secretive and hermetic, even though we Puerto Rican residents have a constitutional right to access information. Historically, human rights activists, environmental leaders, and freedom fighters have been bold, wise, and intuitive enough to find out much of what has been hidden. But transparency is one of the main pillars of a democracy, and the lack of it in Puerto Rico has caused us so much pain. When establishing the great root causes of our current crisis, most people focus on colonialism, which is obvious and irrefutable. But the lack of transparency is not inextricably linked to colonialism, and it has framed our political culture. The paradox is that this happened while both the US and Puerto Rican government officials perpetrated this narrative that presented Puerto Rico as a highly democratic and even model country. This was all such a shameless fraud, as we never created even the minimum basis for a culture of access to information, a culture of transparency, which are, again, vital elements of a democracy.

As an activist, I can say that I spend at least half of my work time fighting for the unknown, all sorts of data and information that the government does not provide. For example, police use-of-force documentation, the number, information, and circumstances of people killed, mutilated, or seriously injured by the police. Requesting and fighting for this information takes a lot of any activist's workload. And we don't even have a mechanism to access information. It is very common to ask an agency for information and not receive anything back or receive very poor, aggregated data. Your only options then are (1) giving up on the data and using whatever other information you can find on your own or (2) suing the agency, which is not something everybody can do. It is way too onerous to make data-based decisions and do advocacy work in Puerto Rico. This is a very basic element of creating a democracy.

MARISOL LEBRÓN. Before the hurricane, there seemed to be widespread dissatisfaction with PROMESA and the imposition of the Fiscal Control Board. All around Puerto Rico, but especially in the San Juan metropolitan area, there were many demonstrations and actions against the proposed solutions to the debt crisis. How would you describe the political climate before María struck? How did the government respond to the surge of activism against the new austerity regime being implemented?

MARI MARI NARVÁEZ. The government responded with a level of repression that I believe we had not seen in a very long time. Decades maybe. Never in my life did I think I would be living under a similar repressive environment as my parents did. Fifteen years ago, I thought all that had been left in the past. I was very wrong. Repression is awful for many reasons, especially the way it discourages people to exercise their freedoms. Very bad things have happened to people who protest during the last few years, especially the young. That's bad and sad and terrible, but a lot of people don't get intimidated and that's also inspiring. What really seems terrifying, however, is the way that the world's capital controls governments and a country's destiny, just like that. In Puerto Rico, both the Puerto Rican and US governments and the Fiscal Control Board went to great lengths to assure that the board's decisions could find a relatively easy execution path. Police repression, as well as their protection of private interests, has been a central part of this plan.

The federal government, as always, has had a leading role in this climate of repression, and if there's any doubt, just look at what they have done with Nina Droz. Droz left her house on May 1, 2017, to protest the violent austerity measures that are killing our country, and almost two years later, she has not returned. She was denied bail and was unjustly incarcerated using an eccentric theory about a "terrorist organization" that they supposedly investigated and is "linked" to this case. A completely crazy theory and everyone knows it. She has been tortured, intimidated, and denied medical attention, and her dignity has been violated continuously. The federal prosecution incarcerated her as a cruel exhibition punishment for all other protesters. And they do it because they know that the only thing that can really threaten their economic exploitation plans are social mobilizations. Protests and protesters can change the course of things.

MARISOL LEBRÓN. How did María change things? Did it make organizing against the debt more difficult? Increase its importance?

MARI MARI NARVÁEZ. It has been a strange and extraordinary time. Some aspects are better now, some are more difficult. The fact that the hurricane unveiled our poverty, our structural and political fragility, has been somewhat positive as more people have been able to understand the mortal effects of austerity measures and the relationship between corruption, capitalist exploitation, the debt, and our vulnerability, our current daily situation. The media attention on Puerto Rico, especially due to Donald Trump and the federal government's discrimination and failure to respond in Puerto Rico, has also brought a new level of philanthropic capital that we did not have before. That has facilitated things for certain NGOs and advocacy organizations, very few of them actually, but most of them are organizations with impact, leverage, and a proven capacity and commitment. That has been positive.

On the other hand, a lot of people are still dealing with the aftermath of the hurricane. And when you don't have a house, for example; when you have an already difficult economic situation that worsens because of the lack of work or the effects of austerity measures; when you're in a bad situation and your children's schools close, and you might have lost your car in the hurricane or maybe you lost your support network because your family moved to the states—you cannot think of mobilizing against the debt or for police accountability or whatever. Many people are immersed in an extreme economic and social situation, and this makes it very hard for them to advocate for their rights. In some instances, advocating for human rights becomes a privilege and that is tragic.

MARISOL LEBRÓN. After the storm, we saw the deployment of military forces to clear debris, engage in "peacekeeping," and help with the distribution of aid. Some people saw this as the militarization of necessary humanitarian relief efforts, while others suggested that the military were the only ones capable of properly distributing aid and keeping the streets safe in the aftermath of María. What was your reaction to seeing this increased military presence?

MARI MARI NARVÁEZ. I think that, unfortunately, we in Puerto Rico are used to seeing the military often. Of course, it was more dramatic in the aftermath of the hurricane. You'd even see them right in the middle of the highway, just standing there in their huge vehicles, sometimes doing absolutely nothing. I'm not super proud about saying this, but I didn't have much of a reaction to the militarization. I guess we're used to that. I think the military presence in the streets didn't feel as awful and overwhelming as it did in Haiti after the earthquake, for example. I was there, and I had a much more visceral reaction to it there than in Puerto Rico.

But those days after the hurricane were very blurry. They were a suspended time. We had so much to do and to worry about, from basic things like finding drinking water to more complicated things like getting to the countryside to bring help or dealing with each other's losses. I wasn't as disturbed with the military presence as I was with the imposed curfew, for example. And, of course, with the general chaos, especially the lack of diesel. Knowing that diesel trucks were arriving at mansions and rich people's buildings but not to the hospitals angered me and freaked me out. You just couldn't believe that was happening. I knew we couldn't afford to end up in a hospital, so I was constantly thinking about how we could stay safe and healthy.

Even when I was very disturbed by the curfew, I didn't do something about it right away. I couldn't. We were engaged in emergency labor and, as I said, dealing with our own personal chaos. But in November, I began to discuss it with different activists and communicators. By December 2017, a group of organizations and activists had already mobilized and created a report on the human rights violations during the emergency management. We had an audience at the Interamerican Commission of Human Rights in Washington, DC. In that report, we analyzed and discussed the curfew, among many other issues. A lawyer who participated convinced me that the curfew had been illegal, and I kept thinking about it. Finally, when the hurricane's anniversary date was arriving, I realized we had to go back and hold the conversation we had been unable to have before: the conversation about the illegal curfew that was imposed on us. At that time, I had already founded Kilómetro 0, and we created an awareness campaign called Mi Candado Lo Tranco Yo (I Lock the Padlock Myself). We brought this message about how fundamental rights such as the right to move had been needlessly violated in our faces, why it was illegal, and why we had to reaffirm to ourselves that "mi candado lo tranco yo." I saw it as a powerful message of self-determination. Yes, we are colonial subjects. Yes, there are many decisions about ourselves that we cannot make. But at the very least, we should all be able to determine when we close our own doors and padlocks. This is a fundamental freedom and basic right of self-determination.

We also highlighted the story of a woman named Ana Luisa Nieto who died because her dialysis center was closed due to the curfew; and the story of Aníbal Martínez Centeno, a man who was illegally arrested at a gas station during the curfew and spent the night in jail.[1] And we underlined the fact that Aníbal was a working-class Black man. We tried to explain why the curfew was illegal and discriminatory, and we did it in the spirit of a conversation that you cannot leave pending, no matter how much time has passed. It was important to say, "OK, at that moment we couldn't do much for so many reasons. But we have to talk about what happened here, about the illegality of this curfew, about how it violated one of our fundamental freedoms and why we cannot let this happen again."

MARISOL LEBRÓN. The storm put on full display the many ways that people and communities find themselves vulnerable to violence and harm in Puerto Rico. Groups that were vulnerable to violence before the storm found themselves even more so after it. There were reports that women and LGBTQ people were experiencing higher rates of harm after María and that the police were largely ineffective in addressing this violence. Why do you think we saw this uptick of violence after the storm, and why has policing failed to protect these groups?

MARI MARI NARVÁEZ. The police have always failed to protect these groups, and that is one of the reasons why they are facing a reform process. These safety problems were fully predictable. Organizations that work with these groups knew that a hurricane-like situation would generate threats against these vulnerable groups, and they anticipated it. It was not a mystery or a surprise. How can you, government, not know that domestic violence victims, for example, would find themselves *incommunicadas* and unable to report their aggressors? How can you not anticipate that shelters needed to have protocols for sexual aggression complaints, for protecting trans people and women's rights? These things are obvious, and the police failed to attend to them because there is no institutional will to do so. Their only vision is the punitive and remedial one. When the crime or the violation is committed, they might step in (if they find the resources to do so). But they had no plan for prevention. And it's not like this was the first hurricane that ever affected Puerto Rico, so there is absolutely no legitimate explanation for these failures.

MARISOL LEBRÓN. There have been reports that the police have harassed and repressed community groups and organizations that stepped in to aid in recovery after María. Why are we starting to see these groups, which were celebrated in the aftermath of the storm, suddenly become targets?

MARI MARI NARVÁEZ. It doesn't surprise me. As I mentioned earlier, there is ample evidence that, in previous years, the police had been accumulating intelligence on demonstrators, political leaders, community leaders, and even lawyers. Those lawyers who participate in the defense of protesters are constantly harassed and persecuted.

Environmental activist Arturo Massol Deyá was selectively arrested in July 2018, and the police fabricated a case against him after having made some demands before a delegation of Democratic congressmen in Adjuntas, Puerto Rico. Massol and the organization he directs, Casa Pueblo, are some of the strongest voices in favor of renewable energy in Puerto Rico and against the gasification plan of the island's energy system after the devastation of Hurricane María.

These are only some of the intimidations directed toward human rights defenders. The police have accessed the documents and personal information of thousands of people who have participated in, commented on, or interacted with certain Facebook pages, through a judicial preservation order in a case of April 27, 2017. The amount of data that they have accumulated and that they can accumulate through the Facebook platform is worrisome, and we all need to be aware of the implications of these new modes of surveillance and their relation to the collection of files on and repression of dissidents.

MARISOL LEBRÓN. Has Hurricane María caused you to shift or rethink how you approach questions of police reform and accountability? In your work with Kilómetro 0, are there new challenges or issues that you're dealing with that you didn't anticipate?

MARI MARI NARVÁEZ. I think the experiences we had during and after the hurricane made us reflect on every aspect of our lives and work. One of the things I've most had to rethink and try to reframe is the labor and emotional conditions in which police officers are working. Police unions have long been reporting that their members have very low morale. Police officers, just like teachers and social workers and so many other workers, have been hit very hard by austerity; many of them are leaving the agency, and others seem to feel very betrayed. I've heard at least one of their leaders say they have done the job of repressing protesters and protecting the Fiscal Board but have not gotten their due. Of course, I think this rationale is noxious, but to understand the consequences of austerity in the different sectors, I have to understand this frustrating way of thinking. This situation concerns me for too many reasons. First, the low morale has different negative effects. On the one hand, many people agree that police officials are dragging their feet. This means they are not performing their job properly. On the other hand, I also suspect they might be less motivated to abide to the new use-of-force policies. It concerns me very deeply that, by mid-February, the police in Puerto Rico had already killed at least four people. At least one of those was an unarmed man, and another one is said to have been unarmed, although the police say he was not. Another one was armed with a machete and the police responded by killing him. To us, that number seems pretty high for such a short time. I get the feeling that they are acting like saving lives is not worth the risk. They shoot first, then find out if the person was armed. This, added to the fact that accountability mechanisms within the agency are not yet put in place, is extremely dangerous for our democratic and equitable aspirations, as well as for our safety.

This austerity narrative has also led most people to accept a questionable discourse of how the lack of police officers is fueling insecurity. It concerns me that even progressive people are assuming this narrative without consulting the data on it or questioning its assumptions. Do we really need so many police officers? Why? The reality is that we used to have a huge police force of almost eighteen thousand. We now have about ten thousand, but we have also lost a lot of population. We are studying the situation, and although we don't yet have a complete analysis, we have not found any truth in the idea that we have too few police officers.

MARISOL LEBRÓN. What does recovery for Puerto Rico mean to you? How do you see that influencing your work moving forward?

MARI MARI NARVÁEZ. For me, recovery is the opportunity to break: break with our past dependencies, with our pervasive culture of inequity. I wish we could take advantage of this terrible situation to acknowledge that we need to build a more horizontal society, a place for everyone to live and work and love in. This is not an idealistic thought. It is an inclusive, compassionate, human rights perspective. Human rights are pretty fundamental, and exercising them is possible.

In the security aspect, I aspire to build a society where we don't need thousands and thousands of police officers. I aspire to a society that approaches safety as something much bigger than just the police or a Taser or firearms. Safety is about health, about a right to safe housing, peace, and tranquility, about women being able to live and walk in the streets without being raped or harassed. It is living in a community whose health and safety is not threatened by coal ash or other environmental contamination. Safety is having access to clean water, to a high-quality public education, access to culture, and access to a job. To accomplish all that, we need to make the state and its police accountable, to keep police repression and excessive use of force in check. In the world that I aspire to live in, institutions do not control people. It is the contrary. People must control their institutions.

1. Mari Mari Narváez, "Mi candado lo tranco yo," Kilómetro 0, September 20, 2018, https://www.kilometro0.org/blog-desde-cero/2018/9/19/ov6mdpl5nknzg3c93lf26dj46iekuc.

AFTERWORD

Critique and Decoloniality in the Face of Crisis, Disaster, and Catastrophe

Nelson Maldonado-Torres

This book offers reflections and creative work that seek to clarify the meaning and significance of Hurricane María, the context in which it took place, and its aftermath in Puerto Rico. While doing so, it uses a trio of related terms that contribute to contemporary Caribbean decolonial thought and to decolonial thought in general. These terms are *crisis*, *disaster*, and *catastrophe*. While these terms are sometimes used interchangeably in descriptions of anticipated or unanticipated devastation, each has a specific meaning that is worth considering.

One reason for exploring in more depth than usual the meanings of *crisis*, *disaster*, and *catastrophe* is that one can obtain a more precise sense of the extent and depth of various forms of devastation and destruction, as well as of different ways of responding to them. One way of responding to events such as Hurricane María is critique. Critique is important because it indicates a shift from considering a crisis, disaster, or catastrophe as a natural event to approaching it instead as connected to human intervention or sociohistorical forces. All the reflections and creative work in this volume share a common view that Hurricane María was not simply a natural event and that understanding it and explaining it require the consideration of ideologies, attitudes, and social, economic, and political systems, among other factors. While these tasks cannot be underestimated, it is doubtful that critique itself, even in more developed forms, provides an appropriate response to Hurricane María, its context, and its aftermath. This volume points to both the importance and the limits of critique while it also calls for decolonial thinking. A closer look at the meanings of crisis, disaster, and catastrophe sheds light on this point.

Crisis, disaster, and catastrophe may not, in fact, be as similar as they seem, and their differences have implications for how we understand the task of Caribbean thinking. There is a history of links between critique, crisis, and Eurocentric understandings of the crisis of modernity that is unsettled by attention to disaster and catastrophe in the Caribbean. One is tempted to say that—at least in dominant approaches—crisis is to Europe what catastrophe is for the Caribbean. This does not mean that there are no crises in the Caribbean, but that any such crisis is likely to be better understood with reference to catastrophe. Similarly, in European descriptions of European history, crisis seems to take over catastrophe. Yet the history of the Caribbean makes clear that Europe cannot so easily disentangle itself from the catastrophe that is the long presence of colonialism, naturalized slavery, extractivism, and their aftermath in the Caribbean. This points to the need for an account of Western modernity as catastrophe instead of, as is more common, in terms of crisis.

Crisis is etymologically related to the concept of critique, which is central in the definition of critical theory. As Reinhart Koselleck explains, both *critique* and *crisis* have roots in the verb *krino* (whose present active infinitive form is *krinein*), which, in classical Greek, has an array of meanings related to the actions of choosing, judging, and deciding.[1] Critique can be understood as the task of making a judgment, as well as an action that makes it necessary to make a decision, or that calls for a decision. Crisis, in turn, refers to a state of affairs that requires a decision because it is no longer stable. Critique brings about crisis, and crisis requires critique—a judgment or a decision. For this reason, it is not difficult to understand how both crisis and critique have been so central in Western modernity. While critique has been posed as "a genuinely modern disposition," crisis has been considered "a genuinely modern and societal and mental state."[2]

In that sense, if this book about Hurricane María were about the critique and crisis of, say, capitalism, or of the relationship between the United States and Puerto Rico, it could have been said that the book showed the "modernity" of its mainly Puerto Rican contributors, and the modernity of fields such as Puerto Rican or Caribbean studies. But Hurricane María was not simply a crisis—it was arguably more like a disaster or a catastrophe, which, upon close examination, are quite different. A crisis is a moment when decision is needed, while in a disaster it is as if a decision has already been taken and the outcome revealed. It is as if the moment of decision has come and gone unnoticed; disaster seems to be the result of fate, as if something went wrong in the universe.

This points to the etymology of *disaster*, which can be easily understood when considering the Italian term *disastro*: it refers to an "ill" or "misguided" star and has also been taken to mean having "ill fortune," as in "a calamity blamed on an unfavorable position of a planet."[3] This connection between disaster and fate indicates why, when disaster happens, the critic appears out of place while the astrologist and the horoscope tend to get more attention. References to the crisis of capitalism, the crisis of the market, the crisis of the humanities, and so on fail to capture the sense of devastation and hopelessness that disaster tends to bring. This could be read as an error in judgment, but it arguably also reveals a truth about the limits of critique and crisis. The more disasters like Hurricane María happen, the more the concept of crisis will continue showing its limits.

Another important consideration is that crisis maintains the sense that something of value can still be rescued, or that something new can emerge out of a dialectical confrontation of terms. Crisis therefore maintains a view of the past, its value, and the possibilities of the present, a view that is not relevant in a context of disaster. Disaster is not as invested in the value of the past, and it puts in question any notion of a productive dialectic. After a disaster, silence, laments, and speculation reign. As a result, a gap opens up within hegemonic discourse, and there emerge provocations to identify and challenge modernity itself, particularly if the disaster in question occurred in a place like the Caribbean, the history of which is entangled with the making and unfolding of Western modernity. This gap and these provocations become all the more explicit in the context of catastrophe.

While there is something to gain by referring to Hurricane María as a disaster instead of merely as a crisis, the most adequate term for it might be *catastrophe*. The difference is important. Claudia Aradau and Rens Van Munster point in the right direction when they argue that "although 'disaster' and 'catastrophe' are often used interchangeably . . . catastrophes appear to bring that undefined extra, an element of 'un-ness' that crisis and disaster do not capture." Echoing Ulrich Beck, they suggest that "catastrophes are incalculable, uncontrollable, and ultimately ungovernable. They do not just seem unmanageable, they *are* unmanageable."[4]

The distinction between disaster and catastrophe started to become more significant after Hurricane Katrina.[5] For the purposes of this reflection (after Hurricane María), the distinction between disaster and catastrophe is relevant to understand the difference between critique and critical theory, on the one hand, and decolonial thought, creation, and praxis, on the other. Here again Aradau and Van Munster's considerations are useful to chart the path:

> The catastrophe to come induces new problematizations and modes of questioning that are related but not reducible to problems of dangers, risks, accidents, crises, emergencies, or disasters. In distinction to these other terms, catastrophe probably captures best the sense of the limit or "tipping point" invoked by an unexpected future that introduces a temporal disruption with the present. Its etymology (as opposed to those of disaster, crisis, or emergency) hints at this sense of rupture, surprise, and novelty.[6]

The etymology of *catastrophe* certainly points in another direction than the etymology of *crisis* and *disaster*. Catastrophe is not about a decision or about fate, but about a dramatic turn of events, a "reversal of what is expected" (as in drama) as well as "an overturning; a sudden end." The root words in Greek are *kata* (down) and *strephein* (turn). Catastrophe is literally an unexpected "downturn" of events in the face of which, as Aradau and Van Munster assert, "expert knowledge needs to tackle its very limit: the unknown."[7] Unlike disaster, which makes one wonder about fate, but like crisis, which calls for a diagnosis, catastrophe calls for thinking; unlike crisis, however, catastrophe challenges all existing cognitive frameworks and "induces new problematizations and modes of questioning," to use Aradau's and Van Munster's words, that are irreducible to critique.

Hurricane María's impact on Puerto Rico is perhaps best understood and theorized as a catastrophe, or at least there are good indications that catastrophe is relevant for this task and that it provides indispensable elements. From the play *¡Ay María!* to the multiple accounts and analyses in this book, one finds the search for voices and frameworks that are not satisfied with the function of critics or astrologists. Hurricane María calls for another kind of discursive (and practical) intervention, and this book contributes much to the search for it. This is evident in the book's structure: it combines a play, a dialogue, and reflections from journalists, writers, visual artists, and curators. Neither simply positive knowledge nor just "criticism" and much less divination . . . There is a pause here for sharing accounts, taking stock, and standing up as witnesses to various dimensions of catastrophe.

What I wish to add here is that a particularity of Hurricane María's catastrophic character is that it is directly entangled with other scales of catastrophe. One can refer to Puerto Rico's economic conditions no longer as in crisis but as catastrophic, under catastrophe, or as reaching a catastrophic stage. The debt is unpayable and has been used as a means to strengthen the colonial condition of Puerto Rico. To be sure, this colonial condition itself can be understood as a catastrophe: 1898 being a "downturn" that has greatly framed the history of Puerto Rico and Puerto Ricans for the last one hundred and twenty years. But the entanglement between Puerto Rico and colonialism did not start in 1898. It goes back to the period of "discovery" and conquest of the New World and to the long sixteenth century, which is at the heart of the formation and constitution of the modern Western world.

The effects of the catastrophe of "discovery" and conquest since the long sixteenth century cannot be underestimated. Consider Simon L. Lewis and Mark A. Maslin's recent account of the Anthropocene, or the age in which humans have become the principal contributors to changing the earth. They argue:

> The 1610 Orbis Spike [the point of reconnection between the Western and Eastern hemispheres in the context of European expeditions and colonization, leading to a new age for the earth] marks the beginning of today's globally interconnected economy and ecology, which set Earth on a new evolutionary trajectory. It also points out to the second transition we identify—from an agricultural to a profit-driven mode of living—being the decisive change in Homo sapiens' relationship with the environment. In narrative terms, the Anthropocene began with the widespread colonialism and slavery: it is a story of how people treat the environment and how people treat each other.[8]

For Lewis and Maslin, the start of the Anthropocene, which also marks the "birth of the modern world," is deeply connected to the "catastrophic loss of Native American life." The Anthropocene is a "new geological epoch [that] is built from slavery and colonialism, enabled by a long-distance financial industry." This results not only in "a new profit-driven mode of living," as Lewis and Maslin assert, but also in a normalization of catastrophe, evident in the form of continued dehumanization, expropriation, slavery (and its aftermaths), and genocide, otherwise known as coloniality.[9]

The story of Puerto Rico cannot be told without reference to Western modern catastrophe and coloniality. Hurricane María was a catastrophic event that, among other things, exposed the vulgarity of Puerto Rico's colonial relationship with the United States. Listening to Donald Trump's inaccurate comparisons between Hurricane Katrina and Hurricane María or his complaints that Puerto Rico was throwing the US budget "out of whack," and watching him throw paper towels to Puerto Ricans in need could not but recall Cornel West's warning that a Trump presidency would be a "neofascist catastrophe." In the same sentence, West characterized a potential Hillary Clinton administration as a "neoliberal disaster,"[10] which points to the connections as well as the differences between the concepts of disaster and catastrophe.

To be sure, colonialism and slavery continue in multiple other forms: indigenous reservations and stolen lands, imprisonment and criminalization of Black men and women, and so on. What appears as catastrophic in modern colonialism is not only the direct colonial relations that have existed at least since the early moments of the New World's "discovery," but also the naturalization of the relationship between colonizer and colonized and the reproduction of this naturalization, not only in the cultures, institutions, and psyches of normative subjects, but also in colonized peoples themselves. As the works of Aníbal Quijano and many others indicate, one can also use the term *coloniality* to refer to this nexus, this matrix of power, knowledge, and way of being.[11]

As I have argued elsewhere, coloniality can be understood as metaphysical, demographic, and environmental catastrophe, that is, as a major "downturn" in the definition of peoples, the environment, and the very basic coordinates of what constitutes a human world.[12] Metaphysical catastrophe takes place at the foundational level of civilizational systems and views of the self. In the metaphysical catastrophe of the modern world, ideas about normative civility and normative subjectivity are built on the presupposition of unsurpassable lines that separate human beings into more and less human. This metaphysically catastrophic world sets a divide, not between the divine and the mundane, but between "civilized" people, "nature" as a resource to be exploited, and those whom Fanon referred to as the *damnés,* or condemned of the earth.[13] The modern world is constituted out of the fabrication of multiple lines of damnation that anchor apartheid and dehumanization as forms of being-in-the-world.

In the metaphysical catastrophe of Western modernity, every major area of conceptualizing existence attains a special meaning or connotation: there is not only cultural difference but also colonial difference; there is not only power but also the coloniality of power; not only knowledge but also the coloniality of knowledge; and so on for being, gender, and other categories.[14] Space, time, and subjectivity are all understood and approached in ways that sustain the lines between the fully human, nature, and the less than human (and non-human), the latter being different from and more terrifying and threatening than nature. This is what explains over five hundred years of colonialism and the deeper catastrophe that affects not only Puerto Rico but also the modern/colonial world.[15]

In short, metaphysical catastrophe offers grounds for and is itself strengthened by demographic and environmental catastrophes. Hurricane María is a catastrophe inseparable from the catastrophe of Puerto Rican colonialism (a colonialism that continues in liberal, conservative, neoliberal, and neofascist times) and the catastrophe of modernity/coloniality. Modernity/coloniality and the Caribbean for that matter are part of a multilayered and interconnected catastrophic reality. In the face of this, neither divination nor critique suffices as a response, even as both may have their place in this context. The "downturn" of modernity/coloniality asks more directly for countercatastrophic turns, or decolonial turns (symbolic, material, epistemic, etc.) that explore the limits of dominant cognitive frameworks and venture to think otherwise. These turns are also strongly present in the Caribbean.[16] I am referring to the creation of activities and performances that help generate people who identify and question catastrophe, and about the production of frameworks that can address these questions in ways that promote decolonization and decoloniality.

The principal claim here is that, unlike the concepts of crisis and disaster, catastrophe calls for "new problematizations and modes of questioning."[17] Applying the concept of catastrophe to Hurricane María introduces the need to account for a "downturn" that is deeper and more profound than any crisis or disaster. It is not accidental that the conditions of Puerto Ricans on the island during and after the hurricane, and what many took as a disappointing response to it by the US government, led to a considerable degree of attention to the relationship between Puerto Rico and the United States in US national media. Many came to know for the first time that Puerto Ricans are born US citizens, which led to the question of how citizens are differentially treated. This question could not be disentangled from the island's colonial status, which became a heated topic of debate (once again) on the island and among many Puerto Ricans in the mainland United States.

The catastrophe of Hurricane María cannot be separated from the catastrophe of US colonialism and of a political debate in Puerto Rico captured by the question of the status of the island. Likewise, the catastrophe of colonialism in Puerto Rico cannot be disentangled from the catastrophic effects of European colonization in the Caribbean and the US presence in the region—from 1898 to today, when the United States is not only colonizing Puerto Rico but also keeping an embargo on Cuba and playing a major role in manufacturing coups in Venezuela. European colonialism in the Caribbean, in turn, was only the start of a worldwide project of colonization and peripheralization under the premise of European superiority supposedly evinced in the act of "discovery" of the New World. To think about Hurricane María as a catastrophe requires that we carefully consider these various layers as well as explore "new problematizations and modes of questioning." Beyond the misdiagnoses and responses of forms of critical thinking and human sciences generated out of a sense of either the superiority or crisis of Western modernity, thinking about catastrophe in the Caribbean leads to countercatastrophic responses such as decolonial thinking and decolonial aesthetics and poetics. There are examples of these in this text.

Countercatastrophic thought and creative work seek to reveal the various layers of catastrophe and show their entanglement. This requires a "shift in the geography of reason," to use the motto of the Caribbean Philosophical Association, that considers the Caribbean's prominent role in forming the modern West and the region's wealth of responses to coloniality.[18] Decolonial thinking requires countercatastrophic explorations of time and the formations of space, within, against, and outside the modern/colonial world. It also entails the investigation of the various forms of subjectivity, subjection, and liberation that have taken place under the catastrophe of modernity/coloniality. This leads to an encompassing form of Puerto Rican and Caribbean thought that defies the strictures of critical theory, traditional philosophy, the human sciences, and area studies. Facing the catastrophe of Hurricane María and its aftermath calls for a significant engagement with Caribbean decolonial thought and decolonial thinking at large.

1. Reinhart Koselleck, "Crisis," trans. Michaela W. Richter, *Journal of the History of Ideas* 67, no. 2 (2006): 358. For a further development of Koselleck's thoughts on critique and crisis see Reinhart Koselleck, *Critique and Crisis: Enlightenment and the Pathogenesis of Modern Society* (Cambridge, MA: MIT Press, 1988).

2. Bo Isenberg, "Critique and Crisis: Reinhart Koselleck's Thesis of the Genesis of Modernity," trans. Emily Rainsford, *Eurozine* (2012): 1.

3. See entry for "disaster" in the Online Etymology Dictionary (https://www.etymonline.com). Also consulted the Oxford Living Dictionary online (https://en.oxforddictionaries.com/definition/disaster).

4. Claudia Aradau and Rens Van Munster, *Politics of Catastrophe: Genealogies of the Unknown* (London: Routledge, 2011), 28–29.

5. For example, see Enrico L. Quarantelli, "Catastrophes Are Different from Disasters: Some Implications for Crisis Planning and Managing Drawn from Katrina," Social Science Research Council (2006), http://understandingkatrina.ssrc.org/Quarantelli/.

6. Aradau and Van Munster, *Politics of Catastrophe*, 2.

7. Aradau and Van Munster, *Politics of Catastrophe*, 6.

8. Simon L. Lewis and Mark A. Maslin, *The Human Planet: How We Created the Anthropocene* (New Haven, CT: Yale University Press, 2018), 13.

9. Lewis and Maslin, *The Human Planet*, 156, 319–320. On coloniality and catastrophe see Nelson Maldonado-Torres, "Outline of Ten Theses on Coloniality and Decoloniality," Frantz Fanon Foundation, October 2016, http://fondation-frantzfanon.com/outline-of-ten-theses-on-coloniality-and-decoloniality/.

10. Kurtis Lee, "Cornel West Endorses Jill Stein and Says She—Not Hillary Clinton—Is the 'Only' Progressive Woman in the Race,'" *L.A. Times*, July 15, 2016. https://www.latimes.com/nation/politics/trailguide/la-na-trail-guide-updates-1468606689-htmlstory.html. See also, Christina Wilkie, "Trump to Puerto Rico: 'You've Thrown Our Budget Out of Whack'," *CNBC*, October 3, 2017. https://www.cnbc.com/2017/10/03/trump-puerto-rico-budget.html.

11. See Aníbal Quijano, "Coloniality and Modernity/Rationality," *Cultural Studies* 21, no. 2–3 (2007): 168–78. See also two open-access special digital issues on the "decolonial turn" in *Transmodernity: Journal of Peripheral Cultural Production of the Luso-Hispanic World* 1, no. 2 (2011); 1, no. 3 (2012). For recent introductions to the topic and identification of relevant authors, see Nelson Maldonado-Torres, "The Decolonial Turn," in *New Approaches to Latin American Studies: Culture and Power*, ed. Juan Poblete (New York: Routledge, 2018), 118–27; Walter Mignolo and Catherine Walsh, *On Decoloniality: Concepts, Analytics, Praxis* (Durham, NC: Duke University Press, 2018).

12. For a more detailed account of coloniality as metaphysical, demographic, and environmental catastrophe, see Nelson Maldonado-Torres, "On Metaphysical Catastrophe, Post-Continental Thought," in *Relational Undercurrents: Contemporary Art of the Caribbean Archipelago*, ed. Tatiana Flores and

Michelle A. Stephens (Los Angeles: Museum of Latin American Art, 2017), 247–59; Maldonado-Torres, "Outline of Ten Theses."

13. Frantz Fanon, *The Wretched of the Earth*, trans. Richard Philcox (New York: Grove, 2004).

14. See, among others, Walter Mignolo and Arturo Escobar, *Globalization and the Decolonial Option* (Durham, NC: Duke University Press, 2010); Mignolo and Walsh, *On Decoloniality*.

15. For a decolonial approach to Puerto Rico, see Ramón Grosfoguel, *Colonial Subjects: Puerto Ricans in a Global Perspective* (Berkeley: University of California Press, 2003). For an elaboration of modernity/coloniality as a unit of analysis, see Walter Mignolo, *Local Histories/Global Designs: Coloniality, Subalternity, and Border Thinking* (Durham, NC: Duke University Press, 2000).

16. See, for instance, Maldonado-Torres, "The Decolonial Turn." Starting with the Haitian Revolution, the Caribbean is featured in the various major moments of the decolonial turn, or decolonial turns, that are described in the text.

17. Aradau and Van Munster, *Politics of Catastrophe*, 2.

18. See Lewis R. Gordon, "From the President of the Caribbean Philosophical Association," *Caribbean Studies* 33, no. 2 (2005): xv–xviii; Lewis R. Gordon, "Shifting the Geography of Reason in an Age of Disciplinary Decadence," *Transmodernity: Journal of Peripheral Cultural Production of the Luso-Hispanic World* 1, no. 2 (2011): 95–103.

ACKNOWLEDGMENTS

This project began as a conference by the same name organized by Yarimar Bonilla at Rutgers, the state university of New Jersey on the one-year anniversary of hurricane María. That conference was made possible by the sponsorship of the Office of the Senior Vice President of Academic Affairs, the Department of Latino and Caribbean Studies, the Global Institute for Research on Women, the Center for Women in the Arts and Humanities, the Committee to Advance Our Common Purposes, and the Rutgers Advanced Institute for Critical Caribbean Studies.[1] We would like to particularly extend our gratitude to Isabel Nazario, associate vice president for strategic initiatives at Rutgers, for having the vision and the commitment necessary to make that event possible and to her entire staff for their hard work and dedication. Without Isabel's commitment to bringing attention to the historical importance of Hurricane María, this volume would not have been conceived.

We would also like to thank both the original participants of the conference and the contributors who joined afterward for taking on the challenge that was posed to them. Writing about the consequences and legacies of Hurricane María a mere year after its devastating impact would be difficult for anyone. The contributors of this book carry the added weight of managing their own personal crucibles and experiences of the storm. At the conference where many of these papers were first presented, participants often had to hold back their tears, carefully manage their shaky voices, and push through waves of emotion in order to attempt to offer insight while still in the midst of their own individual and collective trauma. All of this while offering presentations, and later drafting polished essays, in a language that for many is both foreign and imposed.

For those who presented on that first-year anniversary, as well as those who traveled from around New Jersey, New York, Pennsylvania, and beyond in order to attend, the event was an occasion for collective mourning and solidarity. Even as we complete this volume nearly two years after María, Puerto Rico remains embroiled in crisis, confusion, and turmoil. Most of the traumatic processes experienced since the storm remain unprocessed and unaccounted for. We still do not know the names of those that were lost, the full extent of the damage, the scope of government mismanagement, much less if and when there might be some form of repair and redress. We thank our contributors for completing these essays in the midst of their numerous and multisided efforts to push toward a just recovery and hope that this text can, in some small way, contribute to the larger reckoning that is yet to come.

In addition, we would like to thank the team at Haymarket Books, and in particular Anthony Arnove, for recognizing the importance of this conversation and working with us to speedily bring this volume to print. Lastly, we are grateful to Pablo Morales, Dawn Welles, and Kimberly Roa for their deft assistance in editing, revising, and nurturing these pages.

1. Full details of the conference are available at https://academicaffairs.rutgers
    .edu/pr and videos of the performances and panels are available at https://
    livestream.com/rutgersitv/aftershocks.

CONTRIBUTOR BIOS

Yarimar Bonilla is a professor in the Department of Latino, Africana, and Puerto Rican Studies at Hunter College and the PhD program in Anthropology at the Center of the City University of New York (CUNY). Bonilla is the author of *Non-sovereign Futures: French Caribbean Politics in the Wake of Disaster* (2015) and a frequent columnist in venues such as the *Washington Post,* the *Nation,* and *El Nuevo Día.* Bonilla writes broadly about questions of sovereignty, citizenship, and race in the Americas. In 2018, Bonilla was named a Carnegie Fellow in support of her current book project which examines the political, economic, and social aftermath of Hurricane María in Puerto Rico.

Rima Brusi is an advocate, educator, researcher, and essayist. Formerly a faculty member at the University of Puerto Rico, Brusi is the founder, director, and principal investigator of UPR's Center for University Access, an outreach, research, and advocacy initiative. She is an applied anthropologist at the Education Trust, and she is currently writer-in-residence at the Center for Human Rights and Peace Studies at CUNY–Lehman College.

Mariana Carbonell is a freelance theater director, producer, translator, dramaturg, and choreographer based in San Juan, Puerto Rico. She has worked with numerous stage productions and served as production coordinator and stage manager for Tablado Puertorriqueño and Teatro Breve, among others.

Arcadio Díaz-Quiñones is professor emeritus of Spanish at Princeton University, where he also served as director of the program in Latin American studies. He had previously taught at the Universidad de Puerto Rico, Río Piedras. His publications include an edition of *El prejuicio racial en Puerto Rico*, by Tomás Blanco (1985), and two books of essays: *La memoria rota* (1993) and *El arte de bregar* (2000). He also edited the volume *El Caribe entre imperios* (1997). His book *Sobre los principios: los intelectuales caribeños y la tradición*, was published in Argentina in 2006. *A memória rota*, an anthology of his essays translated into Portuguese, appeared in Brazil in 2016.

Tatiana Flores is an associate professor of Latino and Caribbean studies at Rutgers University–New Brunswick. A specialist in modern and contemporary Latin American art, Flores is the author of *Mexico's Revolutionary Avant-Gardes: From Estridentismo to ¡30-30!* (Yale University Press, 2013), winner of the 2014 Mexico Section of the Latin American Studies Association Humanities Book Prize. Her recently curated exhibitions include *Wrestling with the Image: Caribbean Interventions* (Washington, D.C., 2011), *Disillusions: Gendered Visions of the Caribbean and Its Diasporas* (New York, 2011), and *Medios y ambientes* (Mexico City, 2012). Flores is a member of the editorial board of *ASAP/Journal* and *The Association for the Study of the Arts of the Present* and a regular contributor and adviser to *Art Nexus*. Her second book project in progress focuses on the art and visual culture of Venezuela during the presidency of Hugo Chávez.

Sofía Gallisá Muriente is a visual artist working mainly with video, film, photography, and text. In 2011, she co-founded IndigNación, a Spanish-language multimedia collective born out of the Occupy Wall Street movement, and in 2012 she co-founded Restore the Rock, a non-profit hurricane relief organization dedicated to people-powered recovery after Superstorm Sandy. In 2015, Sofia was awarded an emerging artist grant from TEOR/éTica, Costa Rica, where her work was exhibited in a solo show titled *Buscando La Sombra*. She has exhibited internationally, most recently at ifa-Galerie, Berlin; the Getty PST: LA/LA, California; and Espacio El Dorado, Colombia. Since 2014

Sofia has been co-director of Beta-Local, dedicated to fostering knowledge exchange and transdisciplinary practices in Puerto Rico.

Isar Godreau is a cultural anthropologist and researcher at the Institute for Interdisciplinary Research at the University of Puerto Rico at Cayey, where she directs various institutional initiatives and research projects. She is the author of *Arrancando mitos de raíz*, a guide for anti-racist teaching (2013), and *Scripts of Blackness: Race, Cultural Nationalism and US Colonialism in Puerto Rico* (2016).

Christopher Gregory is a Puerto Rican photographer based in New York City who works extensively on various photographic projects in Puerto Rico and Latin America. His work is particularly interested in examining the residue of political and colonial power. He is a founding member of Blackbox, a visual cooperative that merges the creative processes of photography and design to build immersive stories, and he lectures in the New Media Narratives Program at the International Center of Photography.

Mónica A. Jiménez is a historian and a poet whose research and writing explore the intersections of law, race, and empire in Latin America and the Caribbean with a focus on Puerto Rico. She has received fellowships in support of her work from the Ford Foundation, the Puerto Rican Studies Association, the Institute for Global Law and Policy at Harvard Law School, and the University of Texas at Austin. She is a Canto Mundo Fellow and an assistant professor of African and African diaspora studies at the University of Texas at Austin.

Naomi Klein is a bestselling author and award-winning journalist, is the inaugural Gloria Steinem endowed chair in media, culture, and feminist studies at Rutgers University-New Brunswick. She is senior correspondent for the *Intercept* and contributor at the *Nation* magazine. She is also a Puffin Fellow at the Nation Institute. Her most recent book, *The Battle for Paradise: Puerto Rico Takes on the Disaster Capitalists*, was published in June 2018.

Eduardo Lalo is a novelist, essayist, film director, and photographer. Eduardo's novel *Simone* (2012) won the Rómulo Gallegos Prize in 2013. He publishes columns of literary criticism in publications such as *80 Grados*. He has directed two medium-length films: *donde* (2003) and *La ciudad perdida* (2005). In addition, he has shown his work as a photographer in more than a dozen exhibits.

Marisol LeBrón is an assistant professor in the Department of Mexican American and Latina/o Studies at the University of Texas at Austin. An interdisciplinary scholar working across American studies, Latina/o/x studies, and feminist studies her work focuses on social inequality, policing, violence, and protest movements in Puerto Rico and US communities of color. She is the author of *Policing Life and Death: Race, Violence, and Resistance in Puerto Rico* (University of California Press, 2019) and one of the co-creators of the Puerto Rico Syllabus, a digital resource for understanding the Puerto Rican debt crisis.

Beatriz Llenín Figueroa's research and creative work revolve around Caribbean literature and philosophy, island and archipelagic studies, gender and queer theory, decoloniality, and street theater and performance. She teaches at the UPR-Mayagüez campus as an adjunct faculty member in an adjunct country, where she also works as editor for Editora Educación Emergente and as freelance editor and translator. Through her work with the collectives PROTESTAmos and Taller Libertá, she fights for a decolonial future for the archipelago, debt relief and reparations, public education, and independent art in Puerto Rico. Some of her creative work on Puerto Rico's current crisis has been recently published in *Puerto Islas: crónicas, crisis, amor* (2018).

Hilda Lloréns is a cultural anthropologist and a decolonial scholar. She is the author of *Imaging the Great Puerto Rican Family: Framing Nation, Race, and Gender during the American Century* (2014). The thread that binds Lloréns's scholarship is understanding how racial and gender inequality manifest in cultural production, nation building, access to environmental resources, and exposure to environmental degradation. Lloréns's research has been centrally concerned with critiquing structural inequalities and dismantling taken-for-granted notions of power.

Natasha Lycia Ora Bannan is a human rights lawyer. Her work focuses on the economic exploitation and discrimination against low-wage Latinx immigrant workers, as well as legal support in the face of the economic and humanitarian crisis in Puerto Rico. She works on both domestic litigation and international advocacy before human rights mechanisms concerning issues including: state-sanctioned violence and failure to protect; self-determination and decolonization processes; gender justice; and immigrants' rights. Natasha is associate counsel at LatinoJustice PRLDEF and president of the National Lawyers Guild, the nation's oldest progressive bar association. She is the second-youngest president in the guild's history and its first Latina.

Nelson Maldonado-Torres is a professor of Latino and Caribbean studies, and comparative literature at Rutgers University–New Brunswick. Maldonado-Torres is director of the Rutgers Advanced Institute in Critical Caribbean Studies and the former President of the Caribbean Philosophical Association (2008-2013). He has been selected in 2018-2019 as distinguished visiting scholar by the Academy of Sciences of South Africa. Professor Maldonado-Torres is currently working on two book-length projects: "Fanonian Meditations," exploring the epistemological basis of "ethnic studies," and "Theorizing the Decolonial Turn," which provides a historical and theoretical overview of the "decolonial turn."

Arturo Massol-Deyá is the executive director of Casa Pueblo. He is from the mountainous area of Puerto Rico in the municipality of Adjuntas, where his parents founded the community-based organization Casa Pueblo. Massol-Deyá grew up in this project and has chaired its board of directors since 2007. A graduate of the public school system (1986) and the University of Puerto Rico (1990), he obtained his doctoral degree from the Center for Microbial Ecology at Michigan State University in 1994. Since then he has been a faculty member at the Department of Biology of the UPR Mayagüez campus.

Carla Minet is a journalist and executive director of the Center for Investigative Journalism in Puerto Rico, and for six years she was executive director of the grassroots organization called Prensa Comunitaria (Community Press). Her investigative work ranges from political campaign donations to environmental issues and government affairs. For the past fifteen years she worked as a reporter, researcher, editor, and producer for radio, television, and online, in traditional and independent media outlets such as Channel 6 Public Television, Radio Universidad, *El Nuevo Día*, Univision, *NotiCel*, *80grados*, and *Diálogo*. She has been a speaker at different conferences and forums, a media trainer, and a professor at the University of Puerto Rico.

Sarah Molinari is a PhD candidate in anthropology at the CUNY Graduate Center and a visiting researcher at the Institute of Caribbean Studies at the University of Puerto Rico, Río Piedras. She is also a collaborator on the Puerto Rico Syllabus project and a member of the Puerto Rican Studies Association executive board. Prior to graduate school, Sarah co-directed an oral history project at the Center for Puerto Rican Studies at Hunter College. She is currently conducting ethnographic fieldwork in Puerto Rico on the politics of debt resistance and hurricane recovery with a National Science Foundation grant.

Ed Morales has written for the *Nation*, the *New York Times*, the *Washington Post*, *Rolling Stone*, and the *Guardian*, and was a staff writer at the *Village Voice* and columnist at *Newsday*. He is the author of *Latinx: The New Force in Politics and Culture* (2018), *The Latin Beat* (2003), and *Living in Spanglish* (2002). While a Columbia University Revson Fellow, he produced and co-directed *Whose Barrio?* (2009), an award-winning documentary about the gentrification of East Harlem, inspired by his *New York Times* essay "Spanish Harlem on His Mind." His forthcoming book *Fantasy Island: Colonialism, Exploitation, and the Betrayal of Puerto Rico* will be published in the fall of 2019. Morales is currently a lecturer at Columbia University's Center for the Study of Ethnicity and Race and the CUNY Graduate School of Journalism.

Frances Negrón-Muntaner is a filmmaker, writer, curator, and professor of English and comparative literature at Columbia University. Negrón-Muntaner is the founding curator of the *Latino Arts and Activism Archive* at Columbia University. Among her books and publications are: *Boricua Pop: Puerto Ricans and the Latinization of American Culture* (CHOICE Award, 2004), *The Latino Media Gap* (2014), and *Sovereign Acts: Contesting Colonialism in Native Nations and Latinx America* (2017). She is currently the director of Unpayable Debt, a working group at Columbia University that studies debt regimes in the world. She is a curator of and contributor to NoMoreDebt: Caribbean Syllabus (first and second edition) and director of Valor y Cambio (Value and Change), an ongoing art, storytelling, and community currency project in Puerto Rico.

Patricia Noboa Ortega is an assistant professor of social sciences at the University of Puerto Rico–Cayey. Noboa Ortega is the President of the board of directors of Psychoanalytic Society of Puerto Rico. Her research work focuses on examining factors that mediate Latina women's decisions about their sexuality and sexual health, specifically in Puerto Rico. She collaborated on a project that developed an HIV/STI intervention for Latina women, which was recognized by United Nations AIDS. She has studied psychoanalysis in Quebec, Canada, and has disseminated her research papers in peer-reviewed journals.

Ana Portnoy Brimmer holds a BA and MA in English from the University of Puerto Rico, and is an MFA candidate in creative writing at Rutgers University–Newark. Ana is the inaugural recipient of the Sandra Cisneros Fellowship, a Pushcart Prize nominee, and a co-organizer of the #PoetsForPuertoRico movement.

Erika P. Rodríguez is freelance photographer, and her work has appeared in the *New York Times*, the *Washington Post*, and *CNN*, among other venues. Being from a place that is culturally Latin American but politically a territory of the United States has shaped Rodriguez's interest as a documentary photographer, to explore the topics of community and identity. After Hurricane María ravaged the Caribbean, she was one of the lead photographers covering the aftermath for the international media. Currently based in San Juan, Puerto Rico, Rodriguez covers Puerto Rico's economic crisis, its slow recovery from disaster, and its colonial condition.

Eva Prados Rodríguez is a lawyer and human rights defender. She is a spokesperson for the Broad Movement of Women of Puerto Rico and the Citizen Front for Debt Audit, a collective of organizations and individuals that are demanding a comprehensive and citizen audit of the public debt of Puerto Rico.

Marianne Ramírez-Aponte is executive director and chief curator of the Museo de Arte Contemporáneo de Puerto Rico. Under her tenure the MAC has been awarded the Solidarity in Education Award 2017 granted by the Miranda Foundation. She is cofounder and member of the board of directors of important advocacy and public service organizations such as Puerto Rico Museum Association, Arte Santurce Cultural Alliance, and Movimiento Una Sola Voz.

Carlos Rivera Santana is a decolonial and visual cultural studies scholar. Rivera Santana is currently a research associate at the Center for Puerto Rican Studies at Hunter College–City University of New York (CUNY). For over seven years Rivera Santana was based in Australia, where he completed his PhD in cultural studies and philosophy at the University of Queensland, the university where he became an assistant professor specializing in cultural and postcolonial studies. His forthcoming book entitled *Archaeology of Colonization: From Aesthetics to Biopolitics* will be released in fall 2019 as part of the book series Critical Perspectives in Theory, Culture and Politics.

Giovanni Roberto is a social justice organizer and the director of the Center for Political Development in Puerto Rico, an umbrella organization that set up community kitchens after Hurricane María and now has a number of mutual aid centers around the island. He was a student leader during the 2010–2011 strikes at the University of Puerto Rico.

Sandra Rodríguez Cotto is a seasoned investigative journalist and a public relations consultant, political and media analyst, blogger, and award-winning author. Rodríguez Cotto writes a weekly op-ed column for *NotiCel*, contributes to several media in Latin America and the United States, and writes a blog, *En Blanco y Negro con Sandra*. She hosts a daily radio show on *Red Informativa* and covered Hurricane María on WAPA Radio 680, the only radio network that remained on the air.

Her books include *Frente a los medios* (2014), *En Blanco y Negro con Sandra* (2016); *Bitácora de una transmisión radial* (2018) which received the prestigious Puerto Rican Literature Institute award, and *Mass Media in Puerto Rico* (2019).

Adrian Román is a mixed-media artist focusing primarily on sculpture and drawing. Román's work is informed by issues of race, migration, and identity while exploring both the personal and historical memory of the two disparate worlds that he inhabits: the tropical landscape of Puerto Rico and the overpopulated cityscape of New York. His practice combines drawing, painting, and sculpture within immersive installation environments composed of objects collected from different communities, from salvaged wood and window frames to historic artifacts and vintage photographs. The results draw upon the history and memory embedded in the objects. Román is a member of the NARS foundation and has worked closely with organizations such as Semilla Arts Initiative in Philadelphia and Caribbean Cultural Center African Diaspora in New York. Adrian's work has been exhibited in several solo and group shows in Puerto Rico and the United States.

Raquel Salas Rivera is the 2018–19 poet laureate of Philadelphia and the inaugural recipient of the Ambroggio Prize from the Academy of

American Poets. Their fourth book, *lo terciario/the tertiary*, was on the 2018 National Book Award Longlist, and was selected by Remezcla, *Entropy*, Literary Hub, mitú, Book Riot, and *Publishers Weekly* as one of the best poetry books of 2018. They have received fellowships and residencies from the Sundance Institute, the Kimmel Center for Performing Arts, the Arizona Poetry Center, and CantoMundo.

Richard Santiago (Tiago) earned a BFA from Marist College in New York and a master of fine arts from the Hoffberger School of Painting at the Maryland Institute of Art in Baltimore under the tutelage of Grace Hartigan. He was a university professor at the Escuela de Artes Plásticas y Diseño de Puerto Rico (School of Visual Arts and Design of Puerto Rico). He has worked in many genres, including film, music, sculpture, graphic arts, and street art. Recently, he has focused on creating collective art projects that center on social issues and pivot around the concepts of empathy and solidarity.

Benjamín Torres Gotay studied journalism at Universidad del Sagrado Corazón in San Juan, Puerto Rico. Has worked for various media outlets and for *El Nuevo Día* since 1997, where he is now a special topics writer and columnist. He is the author of the novel *Tatuajes en cuerpo de niña*.

INDEX

ABOUT HAYMARKET BOOKS

Haymarket Books is a radical, independent, nonprofit book publisher based in Chicago.

Our mission is to publish books that contribute to struggles for social and economic justice. We strive to make our books a vibrant and organic part of social movements and the education and development of a critical, engaged, international left.

We take inspiration and courage from our namesakes, the Haymarket martyrs, who gave their lives fighting for a better world. Their 1886 struggle for the eight-hour day—which gave us May Day, the international workers' holiday—reminds workers around the world that ordinary people can organize and struggle for their own liberation. These struggles continue today across the globe—struggles against oppression, exploitation, poverty, and war.

Since our founding in 2001, Haymarket Books has published more than five hundred titles. Radically independent, we seek to drive a wedge into the risk-averse world of corporate book publishing. Our authors include Noam Chomsky, Arundhati Roy, Rebecca Solnit, Angela Y. Davis, Howard Zinn, Amy Goodman, Wallace Shawn, Mike Davis, Winona LaDuke, Ilan Pappé, Richard Wolff, Dave Zirin, Keeanga-Yamahtta Taylor, Nick Turse, Dahr Jamail, David Barsamian, Elizabeth Laird, Amira Hass, Mark Steel, Avi Lewis, Naomi Klein, and Neil Davidson. We are also the trade publishers of the acclaimed Historical Materialism Book Series and of Dispatch Books.